Bringers of Order

Bringers of Order

WEARABLE TECHNOLOGIES AND THE MANUFACTURING OF EVERYDAY LIFE

James N. Gilmore

UNIVERSITY OF CALIFORNIA PRESS

University of California Press
Oakland, California

Cataloging-in-Publication data is on file at the Library of Congress.

ISBN 978-0-520-41013-8 (cloth : alk. paper)
ISBN 978-0-520-41014-5 (pbk. : alk. paper)
ISBN 978-0-520-41015-2 (ebook)

Manufactured in the United States of America

34 33 32 31 30 29 28 27 26 25
10 9 8 7 6 5 4 3 2 1

For Beth

Love is bigger than anything in its way

CONTENTS

ACKNOWLEDGMENTS

This book is only here because of Elizabeth Kaszynski Gilmore. For ten years, Beth has graciously listened to me ramble—often incoherently—as I struggled to figure out this book. She read early versions, and later revisions, and nearly finished versions, all with her rare generosity. When she writes "Bam!" in the margins, I know I've figured it out. This book has traveled with us throughout our years dating in graduate school, the beginnings of our marriage, our early professional careers, and the birth of our twin girls, Jaime and Jordan. I was very near to finishing this book at the end of 2019 when Beth discovered she was pregnant with twins.

And then COVID-19 happened. And our children were born in July 2020. Twins, as you may know, are always classified as a high-risk pregnancy, and as COVID-19 descended and amplified and twisted everyday life around its spread, I knew the next year would change everything. Faced with a need to care for them and to navigate an extraordinarily complex global health crisis while maintaining both of our jobs without any sort of formal childcare, I put this book aside for over a year. I almost did not finish it. It wasn't until late 2021, when I was agonizing over the project's stasis, that Beth encouraged me to just finish it and see what happened.

Truth be told, I finished this book for Jaime and for Jordan. I will leave it to you gracious readers to decide what, if anything, this book contributes to our understanding of technology, everyday life, data, power, and culture. I hope it is something. But more than that, I want my children to know I finished this book. And I did so while always making time to tuck them in at night, play their favorite games, bake cookies with them, and make them laugh.

This book began with a conversation in Ted Striphas's office in February 2014. Torn between several ideas for a seminar paper in Ted's Everyday Life in Theory and Practice graduate seminar at Indiana University, I pitched a half-formed idea about how wearable fitness trackers drew on understandings of "the everyday." Ted's reply, without even a moment to think it over, was "I think you need to write about Fitbit." In the intervening years, Ted has been a champion of the research that formed this book. In many ways, I feel I owe my career to Ted's mentorship.

In Indiana University's communication and culture doctoral program, my early research found an ideal home to grow. I want to especially thank Elizabeth Ellcessor for believing in my work and helping me beyond my PhD years; Stephanie DeBoer for stepping into the role of cochairing my dissertation after Ted moved to University of Colorado Boulder partway into the process; Gregory Waller for asking the hard questions I needed to figure out how to answer; and Barbara Klinger for being, frankly, a role model for a lot of what I was trying to do. More than their extraordinary work as instructors and mentors, the faculty at Indiana showed me how academic work is at its best when it centers kindness and community.

While the faculty at Indiana shaped me intellectually, the other graduate students were my rock. I want to especially thank, in particular and in no particular order, C. J. Reynolds, Dan Hassoun, Alex Svensson, Megan Brown, Forrest Greenwood, Kirstin Wagner, Danny Grinberg, Christopher Miles, Saul Kutnicki, Katherine Lind, Katherine Johnson, Amanda Bates, Lindsey Pullum, Nancy Smith, Iris Bull, Jesse Balzer, and Catalina de Onís. A special place is reserved for Blake Hallinan, whose collaboration I cherish more than words could say.

This book came into view while I worked toward tenure at Clemson University. As a member of Clemson's Department of Communication, I have had the good fortune to be part of a generous and supportive faculty. I want to especially thank Andrew Pyle and Kristen Okamoto for their help as I pushed this book over the finish line. At Clemson, I began collaborating with graduate students and undergraduate students. I want to thank those with whom I share authorship credit on a range of articles: Bailey Troutman, Carla White, Ben Katarzynski, McKinley DuRant, Madeline Hamer, Valerie Erazo, Patrick Hayes, Katherine Kenney, Madeline DePuy, Jessica Engel, Kate Freed, Sydney Campbell, Savannah Garrigan, Cassidy Gruber, Browning Blair, Lindsey Steffen, and Tim Whims. I wish to also thank the advisees with whom I have not (yet) had the opportunity to formally collaborate: Caitlin Lancaster, Kevin Nutt, Erikka Misrahi, Malaysia Barr, Will Nunley, Nathan Locke, David Schaedel, Kendall Phillips, William Seaton, Janeth Sierra-Rivera, and Toni Baraka. The many students in my undergraduate Critical-Cultural Communication course over the years helped me figure out how to actually talk about critical theory, and I worked out chunks of this book through conversations in my Communication Infrastructures graduate seminar. You have all helped me think better and be better.

Sometimes a project just finds the right person. I owe so much to Michelle Lipinski at University of California Press, who seemed to simply "get" this book right away and has been an extraordinary caretaker and shepherd of the manuscript since she first read it. I have told her this an embarrassing number of times, but Michelle's support of my work means the world to me. When I was struggling to finish the book, I turned to Laura Portwood-Stacer's *The Book Proposal Book* for guidance, and I encourage any authors struggling to clarify their book's purpose to do the same. Two anonymous reviewers at UC Press provided precise, important feedback that helped me hone this book's analysis in its final stages.

I have been fortunate to audition parts of this book's argument in other venues. The research in chapter 1 on Oura Ring is derived from "Predicting COVID-19: Wearable Technologies and the Politics of Solutionism," *Cultural Studies* 35, nos. 2–3 (2021): 382–391, copyright Taylor & Francis, www.tandfonline.com/doi/abs/10.1080/09502386.2021.1898021. Parts of chapter 2 are derived from "Design for Everyone: Apple AirPods and the Mediation of Accessibility," *Critical Studies in Media Communication* 36, no. 5 (2019): 482–494, copyright Taylor & Francis, www.tandfonline.com/doi/abs/10.1080/15295036.2019.1658885. I have developed some of these arguments in presentations given at the annual conferences for the National Communication Association and Society for Cinema and Media Studies over the last decade, and I am grateful for the many conversations with colleagues these professional associations have facilitated.

Finally, I would not be here without my parents, Janet and Jerome Gilmore. They gave me a life filled with opportunities for play and for creativity. They modeled how to care about other people, and they always believed in me. I should be so lucky to pass these gifts on to my own children. It was their own fascination with Fitbit in 2013 that first got me thinking about wearables. My dad embarked on a journey to "walk around the world" with his Fitbit; that is, take enough steps and travel enough distance that it would equal what it would take to walk the length of the globe. It took him ten years to do it, and it took me ten years from the moment I first started writing about Fitbits to writing these acknowledgments. I am glad we both reached our finish line.

Introduction

BRINGING ORDER TO LIFE

"THE GOAL OF WEARABLES . . . is to help bring some order to life."[1] This is the ostensibly modest goal Alexander P. Pentland, professor of media arts and practices at MIT, offered for an emergent set of devices at the first Symposium on Wearable Computing, which took place in October 1997 at the Massachusetts Institute of Technology. The event was meant to set an agenda for the development of wearable computers in the decades to come. Researchers, engineers, developers, and journalists convened for a series of roundtables, keynotes, and presentations—complete with master of ceremonies Leonard Nimoy of *Star Trek* fame! The symposium was sponsored by MIT's Media Lab, whose students displayed an array of hardware meant to interact meaningfully with one's body and integrate purposefully into daily life.[2]

While the symposium's presentations cited various limitations—among them size, affordability, battery power, and practicality—to the potential for wearable technologies to diffuse from labs to retail stores in the late 1990s, this moment nevertheless marked the emergence of an *idea*. Wearables, the symposium's contributors seemed to suggest, might prove close to revolutionary, allowing computers to travel with bodies and mediate humans' relationships to their surroundings in new ways.[3] Like many such conferences promoting emergent technologies, the Symposium on Wearable Computing offers a fairly determinist view, suggesting technologies are catalysts of change, solutions to problems, and forces to propel shifts in culture.[4] Behind Pentland's claim that wearables could "bring some order to life" is a suggestion that life needs order to be brought to it. It suggests, however implicitly, that our everyday lives contain chaos and confusion, and computers—compact, wearable, always-on and always-on-us—might work to abate such chaos.[5]

Wearable technologies—small computers that are designed to be affixed to a human body in order to record, track, measure, and/or engage some element of (bodily, daily) routine—come in many shapes and sizes. By the start of the 2020s, consumer-grade wearable technologies (those widely available at most commercial retailers and generally available at price points between $50 and $500, depending on the brand and its capabilities) took many different forms. They included wearable cameras (GoPro); fitness trackers (Fitbit); virtual reality headsets (Oculus); augmented displays (Google Glass); smartwatches (Apple Watch); sound amplification systems (AirPods); and a variety of other sensors, trackers, pads, bands, and straps designed to compute, store, analyze, and/or order elements of our daily lives. The data they produced could, according to their promoters, serve a crucial role in diagnosing illness.[6] They could be entered as legal evidence in murder trials.[7] They could assist with reconstructing fatal accidents.[8] They could help assign life insurance rates.[9] They could allow police officers to record interactions with community members.[10] They could help people work better and more efficiently, while helping managers understand worker behavior.[11]

Many wearable technologies arguably haven't strayed far from Pentland's promise to bring order. An advertisement for the first iteration of the Apple Watch smartwatch in 2015, titled "Rise," demonstrates this quite well.[12] Here, the Watch acts, in turns, as an alarm, a customizable gadget, a smart-home control system, a messaging system, a transit schedule, a QR code scanner, a navigation device, a phone, an exercise tracker, a task reminder, and, ultimately, a timepiece. The commercial ends with the tagline "The Watch Is Here," but as the collage of urban, rural, and domestic locations spanning (at least) North America, Europe, and East Asia imply, the watch can also go everywhere. Not only is the device lightweight and portable, but its wearers can also strap it to their bodies, shifting computational and communicative processes to the top of the human wrist.

The advertising suggests this is a significant boon to how individuals might organize and navigate their everyday routines, imagining ways the watch's interfaces, buzzes, notifications, and metrics assist the overall management of daily life. This is a cluttered commercial, quickly edited to showcase the watch's many possible affordances to help make living in an increasingly fast world more manageable.[13] The commercial demonstrates how, in the language of sociologist Bruno Latour, humans "delegate" a variety of tasks to technologies.[14] Beyond the examples listed in this commercial, some people need to wear such bands to access their place of work, some need them to be monitored by their employer, some need to wear them for physical education courses, some need them to send health information to insurance providers, and some are sentenced to wear ankle monitors.[15] For some, wearable technologies may smoothly integrate into an everyday life as part of pleasurable habits of data gathering; for others, it may heighten existing concerns and fears of monitoring and oppression. These everyday data are treated authoritatively and are used to adjudicate and to arbitrate what goes on in everyday life, such as when videos from police body cameras are released to local publics to make claims about what happened in violent altercations, when wearable heart rate data are used in court courses, or when life insurance companies give policyholders discounts if they reach goals with a Fitbit fitness tracker over a yearlong period.[16]

If you were to take this advertising at face value, you might believe we have entered a new "Age of Order," in which wearables are key players in the use of computational data collection to know more and know better about the conditions of our everyday lives. This book tells a different story about the role of wearable technologies in bringing order. Through attending to various forms of public discourse about and around wearable technologies and their implementation in various domains and practices, this book traces two recurring themes. The first is that wearable technologies are imagined as knowledge-generating devices capable of creating presumably useful data about how people live their lives, and the second is in their implementation, wearable technologies are used as a means of exercising authority over a variety of ways individuals and institutions make sense of human life. This is, of course, a familiar story: new technologies are often heralded with claims that they will solve problems, improve conditions, and facilitate some form of progress, but such claims often mask how technological systems can also produce social problems of their own. Emerging technologies, in other words, can create all sorts of chaos even as their promoters claim they offer means to tame some aspect of everyday life.[17] Many claims to "learn more" about human activities are guises for manufacturing data about the banalities of everyday life to model populational averages, establish norms, set standards, and arbitrate against individuals. Claims to "learn more" often benefit institutions of power as much as they may help any individual. These claims are part of a larger transformation in how daily life is conceptualized, practiced, managed, and surveilled. Wearable technologies participate in a cultural politics around how our technologies and our cultures interrelate—how our ways of doing shape our ways of living, and vice versa—described, at its simplest, as the *datafication of everyday life*.[18]

Datafication has been used as a concept to assess how elements of lived experience are increasingly converted to data of some sort. While this has been an ongoing project for centuries in the building of governments and administration (consider census bureaus, for instance),[19] the push toward explicitly computational forms of datafication aimed at the habits of everyday life has increased over the 2010s and into the 2020s.[20] When wearables are used to produce data about human life and activity, that data can be used—depending on the context and circumstance—to establish standards and norms, to monitor group activities, and to make political and civic decisions. Embraces of datafication as an ideology have contributed to overstated claims regarding how technologies will solve problems and improve daily life.[21] Often, as the chapters in this book chart, these claims are part of public relations efforts to mask the ways technology implementation often increases surveillance and control over some people's work and activity, reinforcing and reinscribing existing forms of marginalization.[22]

I trace these claims primarily through discourse analysis. I am concerned with how institutions or companies position wearable technologies as valuable, the programs they develop around wearables, and the problems and politics of these programs. In other words, I am concerned with not only how wearable technologies become connected to human bodies in their daily lives, but also how they become connected to institutionalized power. Using discourse analysis allows this book to map the cultivation of imaginaries around what datafication is and what it might offer to groups ranging from life insurance providers to police departments to theme park managers.[23] *Bringers of Order* is not necessarily about the processes of domesticating and using technology, nor is it about the processes of design and manufacture in technology industries.[24] While the people building and using technologies are important, and representations of them and some of their public testimonials appear throughout the chapters, this book is invested in tracing the sensemaking processes surrounding the emergence of wearables, especially in the 2010s and into the first years of the 2020s.[25] While many researchers in media studies, science and technology studies, and communication (to name a few areas) are producing research that provides insight into the complicated ways people make sense of and attempt to use wearable technologies to improve their lives, this book is not necessarily about the processes of technology use.[26]

Rather, this book positions the goal of "bringing some order to life" as the attempted imposition of new ways of living that normalize data manufacturing and transform everyday life. Where Michael Pickering once suggested that the etymological links between *experience* and *experiment* clarify how our everyday experiences give us opportunities to learn through trial and error, data-manufacturing technologies like wearables emphasize a more overtly experimental worldview which, among other things, converts everyday life into a laboratory of constant behavioralist experimentation and analysis.[27] Rather than understanding experience as historically, culturally, economically, racially, geographically, sexually (and etc., etc.) variable and multiple, computational data built from lived experience often acts as "a highly prescriptive response to the questions of 'how to live.'"[28] Everyday life is increasingly surveilled as a site of knowledge production and rationalization.[29] Humans are increasingly asked to square their existence against computational processors that infiltrate into an array of bodily functions and routine operations.

Describing the work wearables do as "data manufacturing" emphasizes the act of *producing* data that did not previously exist, rather than conceptualizing data "collection" as the gathering of data that already exist in the world. As Jathan Sadowski has clarified, following Lisa Gitelman, these metaphors are misleading: "Data is not out there waiting to be discovered. . . . Data is a recorded abstraction of the world created and valorized by people using technology."[30] Sadowski also prefers "data manufacturing" to "data mining," as it clarifies the goals of accumulation and its relationship to capitalism. In this way, data manufacturing is distinct from "data extraction," another phrase that is often used in a variety of computational methods to describe the processes of scraping data from sources.[31] The choice of *extraction* resonates with the phrase "data is the new oil," which had some popularity in the first half of the 2010s to describe a rush to suck as much "data" (generally speaking, for all data was, on some level, seen to be valuable) from the proverbial ground of human affairs. In a brief essay published in *Wired* magazine in 2014, for instance, entrepreneur Joris Toonders suggested "Data" (with a capital *D*) had become "an immensely, untapped valuable asset" akin to "oil in the 18th century." This "Data"—which Toonders did not define, but rather spoke of in all-encompassing terms—would create everything from "smooth functionality" to "progress" to "an enterprisewide corporate asset." The idea is most clearly expressed in his claim that "everything can be tested, measured, and improved. If you can measure it, you can improve it."[32]

These claims echo Chris Anderson in his hyperbolic 2008 essay (also published in *Wired*), "The End of Theory: The Data Deluge Makes the Scientific Method Obsolete." As Anderson suggested at the time: "There is a world where massive amounts of data and applied mathematics replace every other tool that might be brought to bear. Out with every theory of human behavior, from linguistics to sociology. Forget taxonomy, ontology, and psychology. Who knows why people do what they do? The point is they do it, and we can track and measure it with unprecedented fidelity. With enough data, the numbers speak for themselves."[33] This "dataist" ideology, as Minna Ruckenstein and Mika Pantzar have described it, was characteristic of *Wired* at the time.[34] It is not difficult to draw a line from Anderson's proclamations to the work companies like Facebook and Netflix have done in the intervening years to build models of human behavior based on data collected from posting or viewing activities.[35] Shoshana Zuboff's theory of "surveillance capitalism" has also emerged as a popular frame for these issues. Zuboff's work also focuses on the extractive nature of technology companies—for her analysis, mostly Google and Facebook—positioning their drive to gather, package, and sell data as not necessarily a behavioral or scientific goal but an economic one. For Zuboff, surveillance capitalism is "the unilateral claiming of private human experience as free raw material for translation into behavioral data. These data are then computed and packaged as prediction products and sold into behavioral futures markets—business customers with a commercial interest in knowing what we will do now, soon, and later."[36] The work of prediction, as Devon Powers has charted, has long been part of advertising, but companies like Google and Facebook have transformed indices of human behavior (e.g., clicks) into data sets that can supposedly divine the future.[37]

Wearables enact these processes through the various sensors, algorithms, and processors embedded inside them, which manufacture data about everything from ambient noise to blood flow. While wearable manufacturers have often framed data manufacturing using the pretext of a more helpful, efficient, and organized way of life, this organization depends on the constant generation of new data about the humans wearing these devices. From camera recordings to ECG measurements, from task monitoring in an office to analysis of physical movements during athletic conditioning, these technologies are not passively recording what was already there, just waiting to be discovered; they are building and producing knowledge.

The foundation of *Bringers of Order*'s argument about the relationships between technology and everyday life can be expressed in the tensions between two quotes. The first, from philosopher and theorist Maurice Blanchot's 1987 essay "Everyday Speech," suggests, "The everyday escapes."[38] For Blanchot, everyday life was defined by its capacity to evade systems of rationalization, for people invent ways of living that resist efforts to colonize their lives. The second, from John Durham Peters's 2015 book *The Marvelous Clouds*, argues, "The history of media is the history of the productive impossibility of capturing what exists."[39] Capture. Escape. Each chapter of *Bringers of Order* explores how wearable technologies are part of efforts to capture, quantify, process, and in some way "know" the mundane and by some accounts ineffable practices of everyday life across the domains of health, accessibility, sport, labor, law enforcement, infrastructure, and ultimately, culture. As mentioned previously, this tension rearticulates "everyday life" into a domain suffused with many kinds of quantitative and computational information bent on manufacturing increasingly granular measures about how people live. This is, in many important ways, hardly anything new.[40] As I gesture to throughout the chapters, the drive to convert life into information has been accelerating since at least the 1800s in the United States and many other places in the Western world. What is important in the present conjuncture is that the institutional uses of wearables throughout the 2010s point to larger transformations unfolding around the epistemologies of everyday life.

To understand wearables as knowledge machines, it is necessary to treat them as *media technologies*. *Media* is a word with many colloquial and professional meanings, from a general shorthand for all forms of journalism to a synonym for *channel* in describing the communication process (how messages travel from sender to receiver through various mediated channels). This book understands *media* more as a series of technologies having to do with the production of knowledge. Think about taking a photograph of your pet. This photograph acts as a record of the pet's existence; it can become a pathway to memories, but it serves to produce knowledge about your pet: if nothing else, knowledge about what they looked like at a given moment in time.[41] Think about watching television news; as scholars of mass media have long explored, news is characterized by things like agenda setting and framing, which set parameters around what viewers think about and potentially how they think about it.[42] Though grounded in a series of examples and case studies, this book's version of media studies is rather existential. I follow Peters's suggestion that "media are crafters of existence."[43] This definition proposes to treat media as constitutive, as productive forces shaping human affairs. Wearables, then, sit in the middle of humans' relationships (in medias res) with a variety of institutions and experience. Peters's line of inquiry follows historians and analysts of media like Friedrich Kittler and Lisa Gitelman, scholars invested in understanding media as systems for recording, storing, and transmitting.[44]

Wearable technologies, while similar in some respects to mobile media such as smartphones, are a particular class of media. Science and technology scholar Sherry Turkle described mobile phones in the 2000s as being constantly in one's hands, pockets, or purses, creating a "tethered self."[45] For Turkle, mobile communication technologies offered a rearticulation of the social collective and different ways of belonging to one another.[46] While wearable devices like the Apple Watch appear to do many of the things phones can do, they are literally—rather than proverbially—tethered. They are "always" on wrists, or strapped to waistbands, chests, faces, or other parts of a body. Virtual reality headsets are necessarily worn over the eyes while in use, whereas identification bands and badges need to be always on one's body to be effective. Wearing is, or at least is imagined to be, a precondition of use. Although there are certainly people for whom mobile technologies are in practice almost always on their person, this speaks to the different mandates for wearable technology administration, which often imagines an ideal wearer who constantly wears a device.

Because of wearables' connections to human bodies, some scholars have studied them through the lens of fashion.[47] Susan Elizabeth Ryan has argued that wearable technologies are part of a longer history of imagining the capacity for technologies to be placed against human bodies.[48] For instance, Ryan's typological category "augmented dress"—which entails a number of context-aware devices like smartwatches—describes "a costume superpower suitably redesigned for the twenty-first century," a century in which traditional elements of life—customs of dress and adornment, for example—are enfolded within informational flows.[49] The notion of an adornment affording the superpower of data manufacture points toward the larger contextual story this book tells about the early twenty-first century; namely, how it extends and amplifies the previous century's administrative concerns about how to trace and govern everyday life.[50]

These connections between wearable technology and fashion point to a toolbox of words demonstrating a linguistic and etymologic overlap. Consider the Latin word *consuetudo*, meaning the act of habituation, from which are derived the English words *costume* and *custom*, among others. Customs are typically thought of as repeatable occurrences that define social mores, and so too are costumes adorned not only for special occasions of make-believe (consider Halloween parties), but also as components of everyday identity construction.[51] Scholarship on performativity has long been attentive to the dimensions of costuming and the ways humans draw on repetitive acts to cement our ways of being in the world.[52] *Adorn* comes from the Latin *ad* + *ornare*, which traces to *ordo*, and its connections to *order*. Adornments are themselves ordering devices. With wearable technology, wearers adorn themselves with habits, a word that again has a double meaning: *habits* is used to mean the robes monks adorn themselves with in monasteries, as well as serving as a synonym for routine.[53] Habit, as meaning both settled tendencies as well as religious garments, demonstrates how the language used to describe everyday life is related to dress.[54] Taken together, these words—costume, habit, adorn—speak to an interrelationship of daily life and garments. They demonstrate the tensions of everyday life that Michael Gardiner has elucidated: that it is at once the domain of the boring and banal, as well as the creative individual.[55] Our costumes are one way in which we mediate ourselves to others.

These media are part of managing daily life in some way or other. Toward the end of the 1500s, *manage* had transformed from a word associated with horsemanship to having the dual meanings of control (especially administrative control, as with a manager) and wielding by hand. The etymological root comes from the Latin *manu* (hand), suggesting that managing is about bringing things to hand. Wearables literalize this, at least to a point: they bring devices to the wrist, to the digit, to various mounted or strapped components that are purported to offer administrative and control capacities for the wearer and/or some sort of supervisor. Manage also shares this etymological root with *manufacture*, which emerged at much the same time from the French, meaning "something made by hand." These words, set into cultural vocabulary nearly five hundred years ago, have resonance today as ways to describe how data are manufactured through a suite of devices humans are asked to "bring to hand."

This section has laid out the various conceptual contours of the book, with a focus on how wearables act as media technologies and how they might be situated within a cultural vocabulary of everyday life. Treating them as knowledge machines recognizes the importance of seeing media technologies as, among other things, means of manufacturing understandings about everyday life. As the following section explores, these knowledge machines are connected to distinct modes of power.

MODES OF POWER

So far, this introduction has explored the central argument that wearables unsettle extant understandings of everyday life through the ways they are intertwined with logics of data manufacture. This section unpacks the three predominant modes of power this book analyzes: normalcy, surveillance, and solution. Each of these is interrelated, and arguably some element of them is present in every case study and set of devices examined in the main chapters. Given their interrelated and mutually reinforcing operations, I want here to briefly outline each and explain how they connect broadly to discourses around wearable technology.

Normalcy

This book borrows its conception of "normalcy" from disability studies, and in particular from Lennard Davis. In *Enforcing Normalcy*, Davis has argued that "disability, as we know the concept, is really a socially driven relation to the body that became relatively organized in the eighteenth and nineteenth centuries."[56] As Davis explored, categories such as normal and abnormal, abled and disabled, needed to be constructed through the development of political, medical, and sociological categories. Even the development of theories like phenomenology depended on relatively clear understandings of an imagined, ideal body.[57] As Davis argued, "The disabled body is a nightmare for the fashionable discourse of theory because that discourse has been limited by the very predilection of the dominant, ableist culture"—disabled bodies, in other words, do not "fit" with how many theories imagine the operations and understandings of human behavior.[58]

Discourses around mainstream technology have often reproduced an arbitrary, normative understanding of bodies. One of the clearest examples of how normalcy is manufactured through wearable technologies is Fitbit's focus on ten thousand steps per day. The popularity of the fitness tracker—and its emphasis on steps taken as a key metric to monitor throughout one's day—led to the emergence of the phrase "get your steps in" to refer to the perceived benefits of choosing to walk while completing various tasks. Even before Fitbit was released, however, scientists were questioning whether accumulating ten thousand steps through measuring pedometer activity in research subjects could guarantee individuals had met physical activity guidelines recommended by organizations like the Centers for Disease Control and Prevention and the US Surgeon General.[59] Writers on Fitbit's public blog have on occasion twisted themselves into a knot to try to make these connections, such as in a 2018 post from self-described "celebrity trainer" Harley Pasternak: "That number [ten thousand steps] may seem arbitrary, but 10,000 steps roughly equates to 5 miles, which (when it includes 30 minutes at a moderate intensity) satisfies CDC guidelines of at least 150 minutes of moderate exercise per week."[60] This single sentence pivots directly from suggesting a number is not arbitrary to explaining an entirely different metric (time of moderately intense activity). Daniel Lieberman has suggested "10,000 steps" started as a marketing ploy for a Japanese pedometer, the Manpo-kei.[61] Other studies produced near the end of the 2010s came to much the same conclusion: that ten thousand steps work well as an arbitrary goal but are not necessary for many kinds of people.[62]

All this is to say that the popularization of "10,000 steps" allowed Fitbit to craft a sense of normalcy around this goal. While it is not necessarily harmful in and of itself to create a general goal for people using a wearable fitness tracker as a means of self-motivation, companies like John Hancock Life Insurance have implemented systems like its Vitality Program, which provides rewards to policyholders based on how often they hit ten thousand steps per day, baking this marketing tactic into the operations of assessing whether insured humans are as "healthy" as they should be (this program is discussed in greater length in chapter 1).[63] This is what it means for normalcy to become enforced via wearable technologies: guidelines on health, productivity, activity, or protocol become observable or manufactured through the technologies and are used to make decisions about whether individuals are behaving "normally" (that is, desirably).[64] This is to say nothing of using Fitbit to gather data for individuals who use wheelchairs or other walking aids, whose gaits or mobility would be considered "non-normative." While Fitbit has ostensibly improved its ability to measure movement over time, recent research on using Fitbit to monitor adults with progressive muscle diseases suggests the devices are only "satisfactory," and the data produced through Fitbit should not be considered synonymous with that from research-grade devices.[65] The discourses around wearable devices risk shifting and constructing perceptions of what is "normal." As scholarship in disability media studies repeatedly emphasizes, building media technologies that can account for and accommodate a range of human experiences is a process and takes ongoing commitment; it cannot be solved by propping up vague values like activity and movement, especially when those can become co-opted into structures serving to further disable and discriminate.[66]

Surveillance

Wearable cameras have been one of the most prominent sites for debates about surveillance. Technology researcher Steve Mann, for instance, has insisted on the capacity for wearable cameras to perform *sousveillance*, or documentation from general citizens rather than those in power, though the results of his work with wearable cameras have been limited.[67] While wearable cameras, such as those worn by many law enforcement officers, are heralded by some as a way to document and reduce violent actions against citizens, surveillance with wearable technologies goes well beyond traditional forms of sight and vision. Consider datafied forms of surveillance, from the location tracking of children through wearable tags and clips to access cards that track how employees move through and use space.[68] These practices are by no means exclusive to wearables and have been part of how mobile media regularly act as means to execute surveillance. Karen Levy, for example, has charted how long-haul truck drivers negotiate data tracking about their driving habits.[69] Amy Adele Hasinoff has analyzed how applications like Life360 are designed to track young people in the name of making them more secure—while also asking them to internalize themselves as surveyed subjects.[70] For Sarah Pink and Veike Fors, the idea of digital copresence—that mobile and wearable devices are, on some level "always with us" throughout our routinized adoption of them—creates new "entanglements" of humans and technologies in self-tracking practices, some of which might lead to increased (self-)surveillance.[71]

While hardly an exhaustive list, these examples indicate how mobile and wearable media technologies are active participants in what David Lyon has called a culture of surveillance.[72] Lyon suggested surveillance has become routinized, making it a largely banal part of daily life for many people. This banality is, of course, political: not caring about surveillance or shrugging it aside as a nonissue ("Why should I care if I have nothing to hide?") refuses to acknowledge how, for many, operations of surveillance extend existing modes of marginalization and discrimination. Simone Browne, for one, has powerfully charted how contemporary, digital modes of surveillance transform historical modes of policing Black bodies.[73] As chapter 3 explores, collegiate athletes become beholden not only to coaches and staff who physically monitor campus activities to determine whether they are actually attending class, but also to newer forms of mobile and wearable devices that monitor sleep to ensure the athlete is not in violation of the program's policies and can remain part of the team (or maintain the scholarship that might afford them access to their classes to begin with).

Wearable technologies are exacerbating devices. They intensify existing regimes of control. Indeed, the surveillance charted throughout this book has much in common with what the philosopher Gilles Deleuze described in the early 1990s as "societies of control."[74] Deleuze, building on the respective critique of his colleagues Michel Foucault and Félix Guattari, noted what he saw as emergent and co-constitutive transformations in technology and capitalism. Deleuze suggested individuals were in the process of becoming *dividuals*. If an individual is "indivisible"—a whole unit, and the functioning unit of, say, census records or voting tallies—*dividual* proposes that an individual can be broken down into many different parts or forms of data. In regard to wearable technology, the production of biometrics like heart rate and blood oxygen and the production of records like steps taken and distance traveled all indicate a drive toward dividing an individual into ever-proliferating data streams that can be produced, gathered, and analyzed.

Solution

Technological solutionism explains how various kinds of people, institutions, and governments believe in the power of technologies in and of themselves to solve social, civic, administrative, or logistical problems. Researcher Evgeny Morozov is often credited with popularizing the phrase in the 2013 book *To Save Everything, Click Here*, which critiques trends of abandoning serious ethical and moral debate about social, political, and cultural issues in favor of developing mechanisms of technological efficiency to purportedly solve such problems.[75] Critics of solutionism are concerned with the framing of technologies as silver bullets, means of killing the proverbial werewolves stalking the countryside of our culture. As chapter 5 of this book charts, the implementation of body-worn cameras by law enforcement officers was originally framed in such logic: the mere presence of a camera would supposedly alter officer behavior for the better, curbing violence and producing responsible officers. The actual story, of course, has been far more complex, with various departments and individual officers manipulating the cameras, forgetting to turn them on, losing footage, refusing to release footage to the public, or engaging in other strategies of obfuscation, which demonstrates how the introduction of these technologies has only produced new problems to solve.[76]

As just one recent example, consider technological solutionist logics during the COVID-19 pandemic. Chapter 1 discusses this in terms of the promotion of wearables that could supposedly predict the onset of COVID-19 symptoms through biometric sensors. As Stefania Milan put it in an exploration of how the pandemic response overinvested in an idea of a "standard human" (which echoes the previous discussion of normalcy), "The calculation exercises we have grown accustomed to with the pandemic often come with unwanted social costs."[77] A number of other researchers, myself included, highlighted this throughout the first year of the COVID-19 pandemic, when an investment in surveillance technologies and wearable biometrics supposedly helped mitigate the viral spread.[78]

Solutionism, especially in moments of public crisis like COVID-19 but also in more banal operations like the architecture of so-called smart cities (as explored in chapter 6 of this book), offers a particular logic through which power can operate. Individuals are encouraged to accept technologies, despite their drawbacks or limitations, *as solutions themselves*. The byproducts of this—discipline, control, and oppression—may only become apparent later. While the phrase technological solutionism has become somewhat popular in the last several years, it is only the latest version of a much larger mode of critique: technological determinism. In determinist modes of thought, technologies are presumed to be the driving force of historical and social change. In determinism, technology causes effects, and those effects are identifiable. As Langdon Winner once argued, determinism depends on two hypotheses: (1) the technical base is taken as a fundamental condition affecting all patterns of social experience, and (2) changes in technology are the single most important source of change.[79] In some academic circles, technological determinism has been assumed to be a logical fallacy. As historian Jill Lepore has argued, "To believe that change is driven by technology, when technology is driven by humans, renders force and power invisible."[80] And indeed, this is what is at stake in unmasking the solutionist claims circulating about wearable technologies: these discourses often reduce the complexity of any given technology, presenting it as a solution in and of itself while masking how things like data manufacturing, surveillance, normalcy, marginalization, or other modes of power might also be at work.

Orientations

Within each of these modes of power are multiple potential user orientations. While not exhaustive, this book's examples reflect five recurring orientations: choice, mandate, pleasure, complicity, and resistance. These are not stable categories, and they overlap and blur in different ways, but they collectively provide a matrix for analyzing how users are related to the three modes of power sketched previously.

Choice is perhaps the easiest orientation to understand. Here, users make a conscious, independent decision to acquire and use a wearable technology. Their decision to put on a smartwatch is voluntary, as is their decision to continue wearing it day in and day out. There are no systems in place that compel them to use these devices. This orientation presumes users have the agency to make informed decisions about technology use. Mandate, conversely, entails any way an institution compels a user to consistently wear a technology, even if they do not want to. A classic example here is ankle monitors, which individuals are forced to wear as part of sentencing agreements in legal proceedings. A person may, for instance, be permitted to remain under house arrest or have some freedom to move within their community, provided their ankle bracelet does not leave a geofenced area.[81] Less obvious, perhaps, are physical education programs that ask teenagers to wear devices that record their activity levels (chapter 3), office workers who are told to wear a badge that helps track their productivity (chapter 4), or collegiate student-athletes who are told to have sleep data sent to their coaches (chapter 3). In these and other instances, users exist within a system of use that may provide some degree of pleasure or utility, but they are not given the choice. Datafication here is part of a protocol of learning about behaviors to render judgment.

Within the dichotomy between choice and mandate, users might also take different positions toward datafication, including pleasure, complicity, and resistance. Those who find pleasure in wearable technology use may or may not understand how these technologies work, but they are able to enjoy their benefits all the same. Social runners who use leaderboards for motivation (chapter 3) and theme park patrons who can access rides and pay for souvenirs through a wearable band (chapter 6) likely feel some sort of pleasure, at least to a point, through technology use. Many of the advertisements and other forms of visual representation this book analyzes establish sensibilities around wearable technology use as enjoyable and personally rewarding. Complicity functions differently and depends on the wearer understanding that datafication has a downside: that the numbers they provide their life insurance company might help *them* lower their premium, while at the same time someone else who is less active may see their premium rise (chapter 1), or continuing to provide data to coaches because it helps a player maintain their spot on a starting roster even as that same form of data collection is used to punish a teammate for missing curfew (chapter 3). Those who are complicit may be aware of the discriminating tendencies of datafication, even witnessing them firsthand, but they either do not care or do not see discrimination as a problem that will impact them—this is not just about surveillance but also about normalcy, and how some users can reap the benefits of technology use if they exist within parameters of normal or expected use. Finally, there are positions of resistance, which describe ways wearers try to cheat the system and get around the expectations of technology use. Police body-worn cameras (chapter 5) are instructive in this regard, as a number of law enforcement officers have resisted the implementation of what they see as a workplace surveillance mechanism.[82] It is important to remember, in other words, that power is rarely a uniform mechanism; there are always some who benefit from and enjoy its operations, just as there are always those for whom modes of power represent further marginalization.

BUILDING A MAP OF TECHNOLOGY AND CULTURE

Bringers of Order is a work of *articulation*, a theory and method of cultural studies that explores the processes through which things come to be connected to each other.[83] Wearables must be understood as fundamentally about the articulation between a device and a body. While the connection is not always or necessarily human—plenty of people strap wearable cameras to their pets, for instance—this book takes an approach that considers the articulation of human body and wearable technology in terms of its goals, effects, and operations across a variety of institutions and domains, pointing to a complicated context of the datafication of everyday life. Stuart Hall described articulation as a unity produced out of difference or the productive effects of two elements coming into connection with one another.[84] It assesses what happens when things come together and how new forms, ideologies, operations, and practices are produced.

Tracing different articulations can help build a map of meaning. In the introduction to *Culture and Society*, cultural historian Raymond Williams suggested that the emergence of five distinct keywords (democracy, culture, art, industry, and class) collectively mapped the emergence of social formation in England from 1780 to 1950.[85] This book is inspired by Williams's idea of assessing relationships across ostensibly different domains to assess changes in our ways of life and analyze public discourse as sites where meaning is constructed and connections are, partly, forged. This book takes the form of mapping, using public discourse to examine how different wearable technologies come to be connected to different institutions and how that connection serves to legitimize the ways these wearables are positioned as ordering devices. These domains are wearable technology and health, wearable technology and accessibility, wearable technology and sport, wearable technology and labor, wearable technology and law enforcement, and wearable technology and infrastructure. Through a variety of examples in these different articulations, I sketch the proverbial geography across which wearables operate, before offering in the conclusion of this book what Stuart Hall and others have called a *conjuncture*, a coming together of forces that can help explain social and cultural formations at particular moments in time.[86] The conclusion insists that we must use the map of meaning the chapters collectively produce to navigate, make do with, and resist the calls to impose order on our lives through the administration of wearable devices.

This map is constructed in pursuit of what Lawrence Grossberg has suggested is one of the aims of cultural studies: telling better stories.[87] That is to say, stories that are not reductive or one dimensional, but try to account for the complexity of relationships that build conjunctures in time and space. To tell a better story about wearable technologies, I argue for a need to tell stories about politics, power, and institutions—forces that are parts of a contextual story about transformations in culture and daily life. It is not so much that wearables make everyday life political—it always has been! It is that these transformations continue to posit data manufacturing as a solution to a variety of supposed problems about the contingencies of daily life, while at the same time institutions seize on this data to measure, surveil, and control populations. We are increasingly asked to cede daily life to Google, Apple, Amazon, Meta, and a variety of other technology companies that continue to consolidate their power. But we are also asked to cede our habits to insurance providers, medical doctors, schools, police departments, work supervisors, and other figures of authority and power who claim, repeatedly, that knowing more is better.

In a similar vein, John Durham Peters has argued for a need to reengage the potential of technological determinism, at least to a point: "To keep denouncing technological determinism in our moment is to risk a mistake graver than granting agency to devices—that of giving up on critique, that is, reflection on conditions of possibility."[88] Peters, drawing from a number of researchers in the 1970s and 1980s who tried to take determinism seriously, reminded us that change *does* come from an accumulation of forces, including technology.[89] While determinism can reductively obscure what's going on behind the scenes or prevent collective discussion of difficult moral, political, and social issues, I share Peters's commitment to telling, as he puts it, "big stories" about technology that acknowledge their role in producing cultural and social change, even as it is always imbricated with other political, legal, civic, and institutional ideologies and practices. In other words, technologies do play a role in shaping our conditions of possibility in given moments of time.

This book is primarily a work of critical inquiry. Its mode is analytical, and its analysis is focused on the politics of wearable technologies as understood through discourses surrounding them. This form of politics is not so much about state governance as about how particular forms of power gain authority within cultural formations in ways that impact the practices of everyday life. As science and technology studies scholar Langdon Winner has argued, "the technological deck has been stacked in advance to favor certain social interests and . . . some people were bound to receive a better hand than others" due to, among other things, cost restrictions, design specifications, normative conceptions of use, education levels, or other barriers to access and implementation.[90] Questioning the relationship between wearables and everyday life reveals larger projects of administration, sedimentation, and control that benefit those who are already understood to be "normal" while continuing to marginalize those who lack power or fall outside the consigned boxes of normalcy.

Methodologically, the book is a critical discourse analysis.[91] It draws on various forms of public discourse and constellates an array of publicly circulating materials, including news reporting, advertising, popular culture texts like film and television, company reports and keynote speeches, legal policies, and terms of use documents. It assesses these forms of discourse in relation to the modes of and orientations to power described previously and critically analyzes discourse around wearables to understand how their implementation facilitates the datafication of everyday life. These materials help chart how wearables are *imagined* (how they are framed and promoted to do particular things) as well as how they are *implemented* (what programs or policies are actually developed to use wearables, and what conflicts emerge around that implementation).

It is also important to clarify what this book is *not*. Because I remain at the level of public discourse and representation, there are no interviews with technology users, and I made the conscious decision not to reach out to some of the individuals discussed throughout these chapters, like the developer of the Humanyze workplace monitoring badge (chapter 4) or law enforcement officials in Greenville, South Carolina (chapter 5). Keeping the focus on publicly circulating material allows the book to analyze how ideologies and imaginaries are built around emerging technologies, how they are interpellated into larger patterns of meaning making. While each chapter takes some time to explain representative technologies and how they work, like Fitbit's step-counting processes, this book is also not a history of design and production. It does not explore the history of these companies, their inner workings, and the processes that led them to build technologies in particular ways. While it analyzes advertising, public presentations, and media interviews, it is focused on the public performances of the wearable technology industry. My emphasis on public discourse is not meant to suggest that qualitative research with technology users or industrial histories of design and production are not as important to study—quite the opposite. While books will hopefully be written in the future that fill those spaces, I see this book's task as using the discourses around wearable technologies to tell a larger contextual story about changing existential conditions and possibilities in the face of the datafication of everyday life.

In drawing my own map of culture across the articulations each chapter explores, I demonstrate that we need to contextualize wearable technologies as important components of the ongoing dominance of datafication. The manufacture of data, occurring through the abovementioned interlocking modes of power, is again not limited exclusively to wearable technologies, but the map this book sketches shows an interlocking system of technological power at work across domains and institutions.[92] To return to the quote that opened this introduction, the project of bringing order to life is hardly innocuous. While many individuals have undoubtedly benefited from self-tracking, data manufacture, recording, and body-sensing mechanisms in many of the devices discussed in this book, there are also many inequities and imbalances that are part of the propagation of these devices. The creation of order need not be the reproduction of existing power. As the conclusion to this book ultimately suggests, we need to emphasize modes of technological engagement that allow us to find individualized joy and marvel at what these devices can do, rather than cede further ground to the manufacturing of our data in the name of helping corporations and other institutions extend the ways they produce knowledge about human behavior.

ORGANIZATION OF THE MAP

To conclude this introduction, I provide an overview of the map of wearable technologies sketched across the chapters. The first two chapters emphasize normalcy, the second two chapters emphasize surveillance, and the third two chapters emphasize solutions. While all these modes of power are, to some level, apparent in every chapter, this demarcation allows for some modes to be emphasized in different parts of the book.

The opening chapter of the book focuses on how consumer-grade wearable technologies were imagined as health-monitoring devices from 2014 to 2022. The chapter is divided into several examples cutting across different devices. The first example focuses on a class action lawsuit brought against Fitbit for an early heart-rate detection wearable that misrepresented its capacity to monitor heart rate. This example shows some of the early failures of heart-rate monitors, as well as the disconnect between advertising and capacity. The chapter contrasts this lawsuit with the incorporation of Fitbits into life insurance policies and how they are used to monitor and evaluate the health of policyholders for setting rates and assessing normative conditions. The second example focuses on Apple Watch's construction of the body at risk through the company's coordination with government agencies and major research universities to suggest the watch is a lifesaving device for mitigating risk of disease or accidental death. The final example analyzes how Oura positioned its smart ring as a personal COVID-19 monitor in the first six months of the pandemic but did not provide peer-reviewed evidence to suggest it could perform this function. Through an analysis of news reporting about Oura, this case study shows how wearables have been positioned as ways to solve, or at least mitigate, a variety of health issues. These examples demonstrate how the development of wearable health-monitoring devices has entailed a process of legitimizing them in coordination with different government agencies, researchers, and corporations.

The second chapter continues the focus on how wearable technologies are related to bodily conditions through examining how in-ear wearables—some of which are called "hearables"—were promoted to personalize the process of hearing. This chapter examines how companies such as Soundhawk and Doppler Labs positioned hearables as devices for improving the ways wearables interact with different sorts of bodies and contrasts them with the release of Apple's AirPods. Taking a disability studies approach to technology, the chapter attends to how these devices are caught between different understandings of the affordance of personalization, with commercials for Apple's devices emphasizing the modulation of musical playback and commercials for Soundhawk emphasizing the modulation of environments and nearby voices. As this chapter argues, Apple's failure to center accessibility features in its advertising reproduces perceptions of able users as ideal users and emphasizes a normative way to understand the affordance of personalization.

Chapter 3 continues the front half of the book's interest in issues of bodies but focuses on devices used to monitor and surveil groups of people. It traces how individuals' records of fitness activity are monitored across runners sharing social data, physical education courses that collect and share student data, and collegiate athletics departments that surveil student performance in training as well as throughout players' days and nights. In physical education classes at the middle- and high-school levels, the Polar 360 system provides constant data monitoring and pits students against each other in ongoing competition. In college athletics, systems like Rise Science track when and how student-athletes sleep. Reading these programs through the lens of surveillance studies, this chapter argues that claims to improve the health of a team or optimize individual players are alibis for entrenching pervasive modes of surveillance into athletic spaces.

Chapter 4 examines how the quantification of athletic performance outlined in chapter 3 has been used as the basis for workplace surveillance wearables. Through an analysis of public remarks from the representatives of the company Humanyze, this chapter examines how a particular logic of work that emerged with worker management in the early twentieth century continues to transform through wearable technologies. In examining the research studies, publications, and promotional materials about Humanyze, this chapter moves away from the focus on physical activity, health, and wellness that characterizes much of the first half of the book and begins to sketch how wearables have attempted to bring order through forms of surveillance in other areas of life.

The final two chapters focus on how wearables are positioned as means for solving civic problems. Chapter 5 examines this through debates around using cameras to try to curb violence by police officers. In 2015 South Carolina became the first state in the United States to pass legislation demanding police departments implement body-worn cameras following the murder of Walter Scott, an unarmed Black man, near Charleston, South Carolina. Through analysis of news reporting in the *Greenville News* over a seven-year period, this chapter demonstrates how body-worn cameras were sold with promises of transparency and accountability. However, because South Carolina law did not allow the camera footage to be obtained through Freedom of Information Act requests, most camera footage has been obscured from public view. Tracing debates about the cameras and the legal apparatus that protects or disseminates their footage, this chapter demonstrates how wearables are imagined as solutions for public issues, such as police brutality and accountability, but their implementation creates an array of other problems and concerns.

Continuing chapter 5's focus on wearables' relationships to public life, chapter 6 examines the implementation of the Walt Disney World Magic Band. Beginning in the mid-2010s, Walt Disney World in Orlando, Florida, allowed patrons to use a wearable band to acquire fast passes to cut the line in rides, purchase food and merchandise, and interact with the park through active and passive sensor-based tracking. This chapter places the Magic Band in conversation with the history of the theme park, particularly its Epcot Center, which Walt Disney designed as a space of experimentation and technological innovation before his death. Magic Band, the chapter argues, carries on this tradition by testing the viability of wearables in semipublic spaces. The data the bands collect are used to help run the park and generate profiles on users, and such strategies are like so-called smart cities. This turn to infrastructure shows how some wearables are being used as projects to literally rewire space and turn it into a site of constant, passive data monitoring. Magic Band is positioned to optimize experiences for patrons within the park, but it largely serves as a means for the park's operators to optimize daily performance based on the data extracted from visitors.

In the conclusion I tie together these different domains of wearable technology implementation, suggesting that they collectively indicate a shift in culture, or a shift in the objects, values, and practices that make up our way of life. While wearables play but one part in the administrative project of datafication, I argue we increasingly live in a cultural moment that prizes an everyday life based on constant monitoring, quantification, datafication, and extraction. The entrenchment of normalcy, surveillance, and solutionism into the fabric of everyday life through technologies like wearables continues to exacerbate extant power structures. When we reproduce the belief that these forms of monitoring lead to healthier lives, or better performance, or more accountable public officials, we reproduce ideologies that also do harm to others who fall outside "normal" health guidelines or whose work performance leads to some sort of punishment or economic penalties, or that extend police surveillance without a reciprocal extension of transparency. Drawing from the application of Deleuze and Guattari's theories in cultural studies research, I position datafication as an assemblage that must be confronted and changed. Exploring what we might do to resist these forms of power, especially when they are implemented in our workplaces, schools, and communities, the conclusion suggests modes of advocacy, literacy, and resistance as ways to begin changing the articulations and offering new sets of choices for how wearable technologies might become useful for more people.

ONE

Health

OR: BRINGING THE HOSPITAL TO THE WRIST

ON SEPTEMBER 5, 2018, Twitter CEO Jack Dorsey appeared before the Senate Intelligence Committee to discuss how his platform was responding to research into how Russian trolls were utilizing the platform to amplify and circulate political discord in the United States. Following the appearance, Dorsey tweeted: "Heart rate during a Senate and House hearing."[1] Accompanying the tweet was a screenshot, ostensibly from the iOS Health App on Dorsey's iPhone, showing his fluctuations in heart rate over the course of the day. This heart rate data had, presumably, been collected from a wearable strapped to Dorsey's wrist, which could passively manufacture his heart rate throughout the hearing. His ability to tweet this information at all relies on a suite of technical devices—not only his platform and the hardware of the Apple iPhone, but also Apple's decision to develop the health app Apple Watch's capacity to collect, aggregate, and store this data on the app. This screen shot relies on the orchestration of different software, mobile communication devices, and wearable technologies. Ostensibly, Dorsey was trying to use his tweet to show how stressed the hearing had made him and how seriously he took it. The hearing caused his heart to perform "abnormally," at least in comparison to its beats per minute on the rest of September 5. Dorsey could provide this publicly as evidence of the intensity of the hearing, with no interpretation or elaboration on his part. Given that studies have repeatedly shown ambivalence about whether consumer-grade wearables like the Apple Watch should be used as acceptable reflections of bodily conditions, Dorsey's aggregation of heart rate statistics rests largely on the cultural values ascribed to this variable.[2]

The capacity to record, represent, and share one's heart rate and other health-related data is just one of the ways wearable technologies produce understandings of health and the operations of human bodies. This chapter examines a variety of ways wearables have become articulated to health monitoring. First, it examines how Fitbit measures steps and heart rate, as well as a lawsuit brought against the company for the advertising claims surrounding its early models of heart rate monitors. From there, the chapter discusses how despite these concerns about accuracy, Fitbits have been used to adjudicate health-related behaviors in life insurance policy programs. These sections use advertising, news reporting, and legal briefs to examine how Fitbit tried to construct attitudes toward its heart rate monitor and how those efforts came into conflict with legal understandings of what the advertising promised. From there, I examine Apple's process of integrating heart rate monitoring, atrial fibrillation detection, and blood oxidation into its Apple Watch models, including how the company legitimized these devices by obtaining Food and Drug Administration (FDA) clearances in the United States for some functionality. This section draws from news reporting and research articles published around the Apple Heart Study as part of public processes of legitimizing the health monitoring features in the Apple Watch. These materials examine Apple's construction of a body at risk that positions the Apple Watch as a necessary device for preventing medical emergencies and death, despite ongoing concerns that the watch's sensors do not work as effectively for some people, such as those with nonwhite skin. Finally, the chapter focuses on how wearables were imagined during the early phases of the COVID-19 pandemic when testing was not widely available, particularly through the Oura Ring. This section analyzes discrepancies between what research scientists said Oura could do and what journalists said it could do, to show how the ongoing desire for wearables to be perceived as legitimate forms of health monitoring allowed for some companies to exaggerate claims and benefit from the chaotic conditions of the COVID-19 pandemic.

This chapter argues that more than just making some forms of health monitoring more accessible for individuals willing and able to pay for these devices, the devices themselves (and the ways they are promoted and utilized) are part of the construction of a body at risk. In the 1980s, sociologist Ulrich Beck developed the concept of a "Risk Society," which understood risk as "a systematic way of dealing with hazards and insecurities induced and introduced by modernization itself."[3] Beck saw phenomena such as atomic plants transforming how humans assessed "risk" on a global and ecological scale. This articulation of wearables to health produces forms of risk calculation, be it through heart rate monitoring, insurance policies, screening for conditions like atrial fibrillation, or viral disease detection. Over the 2010s, risk calculus became an increasing part of discourses around wearables and health and became a prominent feature of wearable technology, such that the act of using a wearable to monitor anything from heart rate to blood oxygen could be framed as an act of understanding one's body as always-already at risk, in need of monitoring to ensure its operations are "normal" and any "abnormalities" can be detected. In other words, these articulations insist on the need to treat human bodies as in need of monitoring while diffusing forms of monitoring traditionally associated with clinics into everyday life.

SITUATING QUANTIFICATION

These devices render one's health quantitatively through data manufacturing. They produce biometric understandings of one's health. Before turning to discourses around these devices, this section examines why the production of biometrics matters, how it is part of a centuries-long struggle to define what counts as empirical data, and how this data manufacturing is part of a diffusion of health monitoring into everyday life. Wearables have offered, at least according to their advertising, the capacity to monitor, screen, and produce a variety of health-related data, most notably heart rate and other issues related to blood flow, such as blood oxidation and signs of atrial fibrillation. This is achieved largely through a process called photoplethsmography, which uses multiple green, red, and infrared LED lights "paired with light-sensitive photodiodes to detect the amount of blood flowing" through a wearer's wrist when a measurement is taken. According to Apple, its watches flash these lights "hundreds of times" per second, allowing them to measure the rate of blood flow and automatically produce a heart rate.[4] In generating these readings throughout a wearer's day and providing summary reports of trends and notifications of outliers, wearables act as compensatory technologies, redressing (at least to a point) the insecurity of health care through assuring a wearer their body is functioning within "normal" range or alerting them to seek care if something appears irregular.

Access to affordable health care has been a substantial political issue in the United States for decades. While the passage of the Affordable Care Act in 2010 ostensibly created pathways for more citizens to obtain health insurance, the costs of health insurance and health care continued to be prohibitive for many, acting as just one of many health(care) disparities across the United States.[5] While these policy debates are not essential to the analysis of health-monitoring wearables in this chapter, suffice to say that health care has remained a concern for American citizens, and it has intensified in its connections to national partisan politics. In this space, consumer-grade wearables like Fitbit and Apple Watch offered a promise not of care, necessarily, but of monitoring and awareness of bodily function. They promised knowledge about present conditions and, eventually, forecasts of potential risks. While a computational device costing several hundred dollars (or more, in some instances) is hardly affordable for many people, positioning it within the larger precarity of health care illustrates one way these technologies are imagined as democratizing and empowering for individuals.[6] Using wearables for health monitoring entails trusting the device to manufacture reliable data as well as instantiating potentially new practices of self-care, in which the constant monitoring and generation of biometric data is part of a supposedly healthy lifestyle.[7]

The articulation of wearables to health extends from self-tracking groups like the Quantified Self (QS) movement. Founded in 2007, this movement is a collection of technology producers and users who believe in the capacity of numbers and counting to lead individuals toward greater self-awareness and self-actualization. Cofounder and technology journalist Gary Wolf helped develop the concept while working for *Wired* in the mid-2000s. Seeing an emergent string of technologies that brought automatic, computerized processes of counting and measuring more into the realm of accessibility and affordability, Wolf extolled the possibilities of self-tracking for habit formation, knowledge production, and personal fulfillment.[8]

QS's motto is simple enough: "self-knowledge through numbers." Participants in QS utilize devices like Fitbit to produce knowledge about their habits and bodies through *n-of-1* studies, in which a single participant is the source of data generation, collection, and analysis (not dissimilar from the qualitative research method of autoethnography).[9] These users ostensibly choose to use wearables for pleasurable reasons. Wolf has drawn on the tradition of amateur science to underscore the power of QS: "The world is full of potential experiments: people experiencing some kind of change in their lives, going on or off a diet, kicking an old habit, making a vow or a promise, going on vacation, switching from incandescent to fluorescent lighting, getting into a fight. These are potential experiments, not real experiments, because typically no data are collected and no hypotheses are formed. But with the abundance of self-tracking tools now on offer, everyday changes can become the material of careful study."[10] This argument comports with James Gleick's claims that the history of information in the latter half of the twentieth century can be largely understood as the gradual ascendance of numerical ways of knowing and its dissemination into all manner of domains and practices.[11] Adherents of QS advocate turning everyday life into a form of "social laboratory," in which various forms of monitoring and tracking turn life into a series of hypotheses and measurements.[12]

Wearables' use of quantification offers possibilities of capture, recording, and analysis to make sense of one's body. Echoing, in an odd way, the philosopher Baruch Spinoza's claim that "we do not even know what a body can do," discourses about wearable health monitors presume they can know quite granularly and comprehensively some metrics of bodily behavior and function.[13] Wearables come to be framed as forecasting, prognosticating, even diagnosing devices, despite concerns from researchers about over-testing for these conditions leading to misdiagnoses and incorrect predictions.[14] With this supposed claim to "know" a body better have come suggestions that health-monitoring devices will revolutionize health care. Cardiologist Eric Topol, for instance, claimed in 2015 that "the future of medicine is in your hands," and that the incorporation of smartphones and other digital tools would turn medicine "upside down" through democratizing how individuals could monitor themselves, understand their bodies, and communicate with doctors through devices, among other things.[15] Topol's suggestion of an impending revolution is in many ways consistent with Wolf's: each saw an emergent articulation of health and technology that would democratize the production of medical knowledge. This knowledge entails an epistemological shift not entirely dissimilar from what Michel Foucault charted in *The Birth of the Clinic*.[16] There, Foucault noted how the development of hospitals and clinics for monitoring sick bodies allowed for shifts in how physicians understood the development of diseases. No longer needing to make house calls or wait until a body had expired to perform an autopsy, medical knowledge could be produced to assist with treating bodies through the standardization of observational knowledge. These discourses around wearables suggest an aspiration to diffuse the clinic into everyday life, where monitoring the body becomes a daily habit.

This monitoring relies on the datafication of bodily functions. Some media and technology scholars have used the phrase "data double" to describe what happens when bodies become reimagined as bits of data, not just atoms and cells.[17] According to Minna Ruckenstein, data doubles "are material to people's lives and a part of knowing and valuing those lives."[18] In other words, data doubles create categories and labels that can organize individuals.[19] Through measuring aspects of life and activity through data collection, individuals can also have their bodily data used as a mode of personal identification.[20] These practices collectively seek to transform elements of a human body into computer information. The biological and the technological, according to digital humanist Btihaj Ajana, "are seen to be 'mediating' each other: the biological 'informs' the technological and the technological 'corporealizes' the biological."[21] Further still, they entangle human bodies in practices of data surveillance, such that celebrations of data accessibility also entrench suggestions that surveillance could be a necessary—even desirable—component of social life.[22]

The suggestion that wearables can assist with monitoring and detecting symptoms for a wide array of conditions, from arrhythmia to atrial fibrillation to COVID-19, indicates a discursive frame in which the technology is offered to allow wearers to interpret their own health status. The interest in health monitoring on which this chapter focuses is not primarily medical, but rather cultural: from the emergence of stethoscopes in the 1800s and polygraphs at the turn of the twentieth century, the accurate measurement of one's heart has been part of a larger claim to access "what's in your heart"—the imaginary that whether someone is lying, or what their values are, can be gleaned from careful measurement of heart rate.[23] The ability for wearers to examine their own heart rate data offers only the latest instantiation of a desire to more intimately and rationally "know" what is going on below the skin. This expansion of quantification, datafication, and health monitoring rearticulates how bodily functions are datafied in the pursuit of a reliable risk calculus.

FITBIT AND THE CALCULATION OF "NORMAL" BODIES

Before the popularization of health monitoring, Fitbit was first sold as a device for counting steps and using those steps to make assessments about movement. Early iterations of Fitbit did not strap to the wrist but were instead small tags that could be clipped to a waistband, shirt, or shoe. Fitbit

One, for instance, had a small screen that would display a wearer's daily step count and distance traveled when a button was pushed. Data from the device could then be synced to a computer or, eventually, a smartphone application. Beginning with the Fitbit Flex in 2013, the company began offering versions of its tracker that could rest on a wrist. When Fitbit moved the fastening of its device from a clip to a strap, this opened up possibilities for additional sensors like a heart rate monitor to be included. As Fitbit began to compete more directly with companies like Apple, it eventually added more touchscreen features and smartwatch functionality. The initial versions of the tracker relied on two different parts that work together to calculate steps: a three-axis accelerometer and a step-counting algorithm. The accelerometer analyzes acceleration data, using multiple dimensions of movement to make determinations about different levels of activity, such as calories burned and distance traveled. The algorithm, by contrast, "is designed to look for motion patterns that are most indicative of people walking. The algorithm determines whether a motion's size is large enough by setting a threshold. If the motion and its subsequent acceleration measurement meet the threshold, the motion will be counting as a step."[24] In other words, the Fitbit does not actually count steps per se. Rather, it combines multiple forms of sensing and computation to create simulations based on carefully modulated deductive reasoning, facilitated through the development of algorithms designed to count steps through, among other things, assumptions about stride length.[25] This deduction produces a data point: one "step." As steps are combined, their frequency and number are processed through additional algorithms, allowing the device and its software to make assumptions about distance traveled (approximately one mile per two thousand steps) and about calories burned (based on the speed at which steps accumulate, along with information one provides the tracker in setting up a profile such as height and weight, which are also factored into the algorithm's equation). The device takes multiple data points and collates them into information the wearer can access either through the device itself (if the device has a screen providing a graphical layout) or on the company's apps or website after the Fitbit syncs with an internet-enabled device via Bluetooth's short-wave radio lengths. Like weather radar and flight simulators, step counts are based on what might be described as a scientifically informed approximation. They come with the knowledge that step counts will never be completely accurate, but this inaccuracy is a necessary trade-off for creating a datafied portrait of bodily movement.[26]

These computational processes depend on normative assumptions about movement that suggest individuals with assistance or wheelchair users will simply "count less" when using these devices. Laboratory studies with the Fitbit have routinely provided mixed results, with some studies indicating it provides satisfactory measurements but still has a variety of measurement errors and "should not be considered an exact step counter, heart rate monitor or calorimeter and Fitbit active minutes are not synonymous with [moderate and vigorous physical activity time]."[27] Step calculations reinscribe normativity, accommodating individuals who match general profiles of ability while introducing statistically significant error rates for others. Fitbit's forums, for instance, contain threads from wheelchair users trying to interpret "step counts" in relation to their arm movements and the ways the device records distance traveled.[28] As Fitbit's implementation of accelerometers and algorithms shows, attempts to codify steps through trackers rely on deductive simulations that might become increasingly close—if never absolutely precise—to capturing the fluid realities of movement.[29]

In 2015 the Fitbit Charge HR added a heart rate monitor to provide another metric of measurement that would allow users to monitor their heart rate constantly and efficiently throughout the day with a quick glance at their wrist.[30] Following public skepticism about the accuracy of this tracker, several research teams studied its accuracy. A study released by Ball State University's Clinical Exercise Physiology Program in early 2016 "found activity trackers to underestimate steps—from 20 to 90 percent—for lifestyle activities."[31] The Ball State researchers also found the Fitbit Charge made "an average heart rate error of 14 percent," sometimes missing up to thirty beats per minute.[32] A separate study from faculty at California State Polytechnic University corroborated these results, finding that "the Fitbit PurePulse Trackers do not provide a valid measure of the users' heart rate and cannot be used to provide a meaningful estimate of a user's heart rate, particularly during moderate to high intensity exercise."[33] After these studies were released, Fitbit issued a statement suggesting its wristbands "are designed to provide meaningful data to our users to help them reach their health and fitness goals, and are not intended to be scientific or medical devices."[34] This statement was perceived to contradict the promise of the device's advertisements: "Know your heart" and "Every beat counts."[35] The "know your heart" campaign was emblematic of a larger shift in fitness trackers in the middle part of the 2010s, as Fitbit and its competitors such as Apple began implementing heart rate trackers to articulate the devices more explicitly to health monitoring.

These devices calculate heart rate via the process of photoplethsmography described previously, in which algorithms and other computational protocols work alongside lights and diodes to calculate heartbeats per minute.[36]

In January 2016 Fitbit users from California, Colorado, and Wisconsin filed a class-action lawsuit against the company, arguing that the devices were significantly incorrect in reporting heart rate measurements, in effect preventing them from knowing their hearts. The lawsuit alleged fraud and misrepresentation, arguing in a press release, "Not only are accurate heart readings important for those engaging in fitness, they can be critical to the health and well-being of people whose medical conditions require them to maintain (or not exceed) a certain heart rate."[37] In June 2018 a judge denied Fitbit's motion to dismiss the lawsuit. As the court noted in its order: "According to the complaint, the ability to record heart rate in real time and during physical activity is marketed as a key feature of the PurePulse devices, yet in reality the products frequently fail to record any heart rate at all or provide highly inaccurate readings . . . Those facts indicate that the devices lack even a basic degree of fitness for use as exercise or activity monitors."[38] The advertised claim to "know" elements of wearers' bodies was at odds with the actual technical function of the device. Fitbit's response was to try to drag a plaintiff, Kate McLellan, into a legal quagmire around arbitration clauses.[39] The following month, United States District Judge James Donato argued Fitbit had acted in bad faith, awarding fees and costs to McLellan while admonishing Fitbit for its conduct and basically warning the company to not try to play legal games with consumers in the future.[40]

Fitbit's lawsuit over this manufacturing of heart rate data demonstrates how discourses construct wearables' possibilities in ways that are not aligned with how they actually function. While that may be true of all advertising, the imaginary of Fitbits as capable of producing knowledge has led to the devices being implemented in life insurance practices built around assessing risk. As historian Dan Bouk has argued, the life insurance industry emerged alongside the rise in statistical capacities for quantifying and measuring individual bodies and groups.[41] Fitness trackers allow for monitoring metrics and using them to build risk assessments of individuals that can impact the economics of their care. Fitbit data interpreted in this way suggests that any human body is always-already at risk, functioning as a constant reminder that the wearer is at risk of being at risk and giving the insurance company new sets of risk metrics. Here, the logic of the at-risk body introduced at the beginning of the chapter returns. Ulrich Beck noted technology's potential for "discovering, administering, acknowledging, avoiding, or concealing" hazards of risk management.[42] Fitness trackers here act as part of a broader system of administering risk management, supposedly helping policyholders become more "financially viable" for insurers. This feeds into neoliberal forms of responsibilization, in which individuals are expected to manage risks and practice adequate self-care.[43] Under these conditions, self-monitoring can be positioned as both desirable and necessary.

Although much of Fitbit's advertising typically targets individuals, its website has a section devoted to "Fitness Corporate Wellness Solutions," a program the company began in 2014 to tout the technology as a solution to motivating employees and ensuring employed bodies can remain cost efficient for insurance providers and employers.[44] Fitbit's early corporate customers included Adobe, BP, GNBC, Redbox, Beaumont, and Boston College, each of which incorporated the devices into its respective wellness program to both incentivize employees toward fitness and monitor their activity through examining their daily routines. Fitbit has also partnered with so-called wellness vendors like Vitality, Cerner, Health Enhancement Systems, Staywell, and Optum, which help companies develop wellness programs for their employees. In January 2017 Fitbit announced a partnership with UnitedHealthcare to develop a wellness program called Motion. By wearing the device, agreeing to have their data synced to UnitedHealthcare's cloud-based server, and reaching prescribed fitness goals, users are rewarded by the program with up to four dollars a day in health-care credits for out-of-pocket expenses.[45] Such a program, however, continues to disadvantage individuals who may—precisely because of health reasons and a variety of preexisting conditions—struggle to meet prescribed goals and earn the very health-care credits that could help with expenses. This system is designed to motivate and incentivize fitness through exchanging data for credit, but it simultaneously marginalizes bodies for whom such a program is incompatible with their ability.

In 2018 the company launched Fitbit Care, which is designed to incorporate "Fitbit's personalized health experience into your health interventions," allowing employers to contract with Fitbit to provide devices and software access to employees in a gambit to improve their physical well-being.[46] At the same time, however, the program is also a surveillance system, which allows "real-time reporting and analytics" to "provide insights to help determine the efficacy of your program, so you can refine and optimize along the way."[47] One such method is to identify what is described as "low utilization" and provide "solutions to further drive participation."[48] Chapter 4 discusses workplace surveillance in greater detail, but here Fitbit pitches itself not as a tool for individual health monitoring, but as a way for employers to track activity and promote competition through fitness goals.

One of the most publicly touted examples of this was insurance company John Hancock's partnership with Vitality, which offered new policyholders a free Fitbit in exchange for providing John Hancock with access to some of their Fitbit data.[49] By logging information through Fitbit and other applications, policyholders can earn "Vitality Points," eventually saving up to 15 percent off their annual premiums. This incentive also implies there is a punishment for nonparticipation, which results in a loss of economic capital for the policyholder who does not provide Fitbit data to the company. In September 2018 John Hancock advanced its integration with Fitbit by announcing it would "stop underwriting traditional life insurance and instead sell only interactive policies that track fitness and health data through wearable devices and smartphones."[50] The program is built around rewards systems and incentives. For instance, Vitality PLUS members can purchase an Apple Watch "for an initial payment of just $25 plus tax. . . . Simply exercise regularly to earn Vitality Points and see the cost of your Apple Watch go down. The more Vitality Points you earn, the less you'll pay each month."[51] In a twist on systems of credit and deficit spending, participants in the Vitality program are asked to trade their body's manufactured data for money, accumulating points to buy fitness technology on credit. Failure to do so each month requires the wearer to pay up to $15.[52] If policyholders would prefer not to pay off a high-end Apple Watch, Hancock provides a free Fitbit with the purchase of a policy.

The Vitality program rewards policyholders with access to different levels of "status." For example, "If you achieve Gold or Platinum status in any year, the premium you pay in your next policy year will decrease compared to the amount you pay in the current year."[53] When John Hancock announced its plan, company CEO Marianne Harrison said, "we are proud to become the only life insurance company to fully embrace behavioral-based wellness," but this might be better described as data-based wellness.[54] It repurposes fantasies of using data to change human behavior, what are sometimes described as "nudges" toward different habits and routines but are often coercive attempts at behavioral modification to align with what an institution deems to be appropriate and normal.[55] The company has claimed that after three years of piloting the Vitality program, it has "logged more than three million healthy activities including walking, swimming, and biking," and policy-holders "engage[d] with the program approximately 576 times per year."[56] Policyholders who use the Vitality program have little choice but to engage the program multiple times per day in order to provide the system with the data to build toward the Vitality Points required to level up to the next status and earn discounts on insurance premiums. In a *New York Times* profile on the program, University of Pennsylvania professor Katherine L. Milkman suggested, "You're going to get healthier people—that's why they're going to give you all their data. . . . The unhealthy ones are going to go other insurers."[57] Milkman's argument repurposes one of the often deployed tropes of privacy debates in surveillance: that those who have nothing to hide need not be concerned.[58] Suggesting healthy people need not be concerned lays bare one of the major goals of such a project: to ferret out "the unhealthy ones" and ensure John Hancock is left with only an insured pool of "vital" bodies who will cost the company less money.

The Vitality Points system works through providing an arbitrary set of points for individuals who meet goals. At an annual health screening, for instance, having a body mass index (BMI) between 18.5 and 24.9 provides a policyholder with one thousand Vitality Points, but having a BMI between 25 and 28 only nets five hundred Vitality Points.[59] Throughout each day, workouts can be used to gain points through syncing Fitbits to the Vitality program. A "light workout," which equals "using your Fitbit device or other wearable device for 5,000–9,999 steps per day" earns ten points. An "Advanced Workout," which earns the wearer thirty points, can be attained through recording over fifteen thousand steps per day. Performing in an event, such as a sponsored run, can earn the policyholder up to five hundred Vitality Points if they "run or walk 12.4 miles or more," but only if they submit a required "proof of event completion" to the company.[60] In developing point systems and rewards out of flawed metrics such as BMI, the Vitality program explicitly rewards those who are willing to produce and provide data.

The Vitality Points program trades an annual review for being constantly inundated with reporting wearable-generated data to the company. This is economically advantageous for John Hancock, in that it helps to develop a system wherein people are supposedly healthier (and thus won't die) because of constantly engaging this system. However, the company's reporting of trends culled from the first three years of piloting Vitality suggests it is also looking for populational averages about how people engage with the program and how to use this data to define what constitutes a healthy or risk-averse body over time. While these programs may not be mandated, insofar as other life insurance companies do currently exist for those who wish to obtain a policy, they demonstrate the user position of *complicity* as described in the introduction. Participants choose to engage and may do so because of the pleasure it provides them, but their involvement in the program and the data they provide can always be used to adjust the population-level understanding of what is average and normal, which may be used to negatively affect others' insurance rates.

Over time, this logic suggests, datafication of health activity becomes accumulated to the point where bodily breakdown and illness might become predictive, and those predictions can fuel policy adjustments for individuals and groups. All that is required, of course, is to constantly tether the devices against one's body and allow for the ongoing data collection and third-party analysis. The broader administrative project of the articulation between wearables and health might be framed here as a desire to use datafication as simultaneously a treatment method, an economic structure, and a monitoring system. These are, importantly, not separate concepts: they are, through programs like Vitality, deeply interrelated.

APPLE WATCH AND THE MONITORING OF THE AT-RISK BODY

When the Apple Watch was first released in 2015, the company seemed to struggle to define what, exactly, it would offer. Gold-plated versions suggested a luxury item,[61] while "taptic engines" supposed different modes of social engagement,[62] and fitness-tracking capacities positioned it as a serious competitor to the Fitbit and its ilk.[63] While the Apple Watch has retained and improved features like messaging and phone calls from the device, the advertising has shifted more toward health screening and emergency response, reflecting Apple's commitment to build legitimacy around the wearable as a serious monitoring device. This section shows how the positioning of the Apple Watch maneuvered away from a general fitness tracker (although it still promotes exercise tracking and daily activity tracking as mainstay features) into a health communication device, one that aspired to serve as an expert on a variety of conditions inside the wearer's body. The overall look of the Apple Watch has not changed much in its first decade. Its front is a square touchscreen interface, with a side button and a small crown assisting with navigating different features on the watch. The underside of the watch—the part that presses against the skin—contains a sensor array that helps enable the heart rate monitor, ECG app, and blood oxygen monitoring. From the first iteration of the device, Apple has promoted the interactivity of its watch, including not only the operations of its touchscreen for performing tasks like sending text messages, but also the complex ways it can tap against wearers' wrists for different sorts of notifications.[64]

Apple devoted significant resources to receiving FDA clearance for the fall detection, electrocardiogram, and irregular heart rhythm detection features of its Apple Watch Series 4.[65] These clearances formally classified the Apple Watch as a "Class II" medical device. While the clearance acknowledged the limitations of the device, noting it is "not intended for clinical use or as the basis for diagnosis or treatment," it nevertheless legitimated the Apple Watch in a way that contrasts directly with Fitbit's class-action lawsuit.[66] Where Fitbit struggled to defend its heart rate monitor, Apple's government endorsement increased perception that the device could manufacture some degree of understanding about whether an individual's body was at risk of becoming at risk. While this form of risk assessment is different from the insurance policies described previously, it still reinforces an articulation of wearables and health based on shifting understandings of risk.

Following the FDA's approval of Apple's ECG feature, several large-scale studies tried to assess the device's accuracy for detecting cardiovascular issues, predominantly atrial fibrillation. Perhaps none of these was as publicized as the Apple Heart Study conducted through Stanford University. Apple formally sponsored this research, which sought to "determine whether software on the Apple Watch could use data from the Watch's heart-rate pulse sensor to identify atrial fibrillation."[67] Enrolling four hundred thousand participants across eight months, the researchers determined that the watch's "pulse detection algorithm has an 84% positive prediction value," through examining which participants received irregular pulse notifications and whether those notifications correlated to individuals who were found to have atrial fibrillation through additional testing.[68] The results of the heart study, which Apple enthusiastically promoted on its main website,[69] were discussed in various national news outlets. *Forbes* journalist John Koetsier led his report on the study with the following interpretation: "First, the obvious: wearable tech seems to make us safer."[70] Other outlets, like *CNBC* and the *Verge*, used considerably more hedging language, incorporating cardiologist commentary or references to previous studies to suggest this study (while promising) should not lead anyone to treat the watch as a diagnostic tool.[71] The Stanford Heart Study nevertheless gave Apple more legitimacy in framing its device as a health-monitoring tool.

With the ostensible support of academic researchers and the federal government, Apple more forcefully framed the idea that wearers can delegate the task of monitoring elements of their heart, blood, and other bodily functions to the Apple Watch. This manifested most explicitly in the campaign "Dear Apple," which featured actors recreating and reading aloud letters that had been sent to Apple CEO Tim Cook describing how the Apple Watch changed their lives. The 2017 version of this campaign, released one year before the FDA clearance, emphasizes fitness and activity. The testimonials in the commercial describe individuals trying to close their activity rings, finding time to do exercise more frequently, and feeling motivated to lose weight. Two letters focus on life-saving capacities, one describing using the SOS feature to call emergency dispatch after a car accident and another describing how the watch's notification encouraged them to seek medical attention.[72] The commercial embodies what Julie Passanante Elman has described as the cultural politics of disability surrounding fitness trackers, in which inspirational messaging encourages potential wearers to use the device to make their bodies more "normal."[73] It deploys inspirational rhetoric of individuals striving to overcome supposed obstacles *through*, *alongside*, and *because of* their decision to wear an Apple Watch.

Contrast this with the 2022 version of the Dear Apple campaign, in which individuals describe harrowing instances like accidentally falling into a trash compactor, struggling to breathe, falling into a frozen river, tracking their erratic pulse while sitting at home, encountering a bear in a cabin, and surviving a plane crash in the Vermont wilderness. These stories are considerably more dramatic, each emphasizing the need to use the watch to call emergency responders or receiving notifications that inspired the wearer to go to the hospital. Multiple testimonials in this commercial contain variations of the phrase, "The Apple Watch saved my life."[74] While my goal here is not to question whether these stories are true (and, full disclosure, I know someone whose life *was* indeed saved because their watch detected a fall and automatically called emergency services), these commercials express a logic around the perceived need for individuals to wear these technologies, especially those who are already classified as at risk, such as elderly individuals more at risk of falling. Apple has worked to reposition the watch from being "just" a fitness tracker to also being a lifesaving device all individuals ought to wear regardless of their condition. This logic is like the life insurance policies previously discussed, but the risk has less to do with capital and policy writing and more to do with the need for wearers to perceive how this technology emphasizes the necessity of monitoring throughout everyday life. This is coercive in a perhaps subtler way than the life insurance programs, suggesting that unless one dutifully wears an Apple Watch they may not know they need to go to the hospital for emergency medical treatment, or they may not be able to reach help when they need it. As a final example to consider public discourse and around the articulation between wearables and health, the following section turns to the promotion of smart rings to supposedly detect the onset of COVID-19 in 2020.

OURA RING AND THE MISREPRESENTATION OF HEALTH MONITORING

In the first year of the COVID-19 pandemic, widespread community testing was not available in the United States, and rapid antigen tests had not yet been made available to the public. In April 2020 a feature in *Wired* magazine asked: "Can a wearable detect Covid-19 before symptoms appear?" This question introduced a glowing review of researchers trying to use sensors embedded in wearable technology to "help track the onset of infections or illness."[75] Despite the leading question, the answer the article offered was: probably not. While there was no peer-reviewed research or reviewable public data to support any claims that wearables would be important tools in the fight to contain the spread of COVID-19, the promise that wearables could predict disease symptoms flourished throughout the spring and summer of 2020. In June 2020 the National Basketball Association (NBA) announced it would be convening an abbreviated season inside a manufactured "bubble" in Orlando, Florida.[76] Apart from constant COVID-19 testing, the NBA also announced the purchase of over two thousand Oura "smart rings."

While some researchers were using Oura Rings to attempt to continuously track whether a wearer was developing fever-like symptoms—and in turn developing COVID-19[77]—the mere existence of this research led to the repeated assertion that Oura Rings could predict COVID-19 symptoms in individuals.[78] Oura's imagined solution to the problem of predicting COVID-19—and the public failing of the US government to quickly develop community testing procedures—fit the pattern of how emergent technologies signal supposedly innovative means of solving public crises.[79] As this section argues, Oura did not so much solve as actually help propagate the crisis in knowledge production about COVID-19 infections. Reproducing promotional language in news outlets embraced assumptions about how one goes about forming knowledge about health and potential susceptibility to COVID-19.

As with the use of Apple Watch to screen for potential conditions such as atrial fibrillation, "prediction" has been a persistent trope in how new technologies are framed.[80] The outward appearance of the Oura Ring is a slightly thick but nonetheless conventional ring. It is meant to fit on a human finger, so ordering one entails providing Oura with a wearer's ring size to ensure it fits appropriately. The inside of the band is lined with three sensors used to manufacture the data Oura processes to create its "Readiness Score." These include infrared photoplethysmography sensors to measure heart and respiration, a negative temperature coefficient sensor to measure body temperature, and a three-dimensional accelerometer to measure movement.[81] The ring is a passive data producer; wearers cannot interact with it apart from choosing to take it off and must check the dedicated smartphone application to receive reports on bodily function and how Oura interprets their "readiness." Oura Rings construct biometrics using the band's sensors—including heart rate variability and body temperature—and employ machine learning software to build correlations for predicting one's wellness in the form of a "score." Oura's "Readiness Score," available on an accompanying smartphone application, monitors "signals from your body and picks up on daily habits to determine how well-rested you are."[82] Oura's website praises the device's tracking capacity: "While Oura is not a medical device, its capabilities are near perfect when compared to advanced medical technologies," including claims of a 98.4 percent reliability rating.[83] The sales pitch to be near perfect is qualified, as one might expect, in the "Precautions" section of the device's terms and conditions document: "Oura Services are not intended to diagnose . . . any disease or medical condition."[84] The advertising here—so close to a medical device that it might as well be one—amplifies Oura's claims about the veracity of prediction, but these documents make clear this datafication is not actually medical in nature. Oura's capacity to understand and predict one's readiness rests on its algorithm's ability to assess ongoing trends in the datafication of a wearer's vital signs, sleep habits, and activity levels.[85]

Many of Oura's claims to predict individual COVID-19 symptoms came from both anecdotal evidence of Oura users and preliminary results of a study based at West Virginia University (WVU)'s Rockefeller Neuroscience Institute (RNI). The WVU researchers, who formally partnered with Oura, developed a separate smartphone app that "goes beyond physical symptoms and body temperature tracking through a holistic integrated neuroscience approach—measuring daily changes in physiological, psychological, cognitive, and behavioral biometrics."[86] The app reportedly allows "the data analytic team . . . to predict the onset of physical symptoms before they occur." To do this, passive data collected from multiple wearables (including heart rate variability and body temperature) are integrated with user-submitted data (survey questions about fatigue, attention, memory, exposure to COVID-19, sense of smell, and other factors) in a smartphone app. These data are then combined and analyzed in the RNI cloud to forecast the likelihood of individuals having contracted the virus, as well as model outbreaks and recovery periods across participants.[87]

While Oura is part of this "holistic and integrated neuroscience platform," it is, importantly, only one component.[88] A press release on May 28, 2020, which was heavily cited in subsequent reporting, claimed "the RNI has created a digital platform that can detect COVID-19 related symptoms up to three days before they show up," but again, there were no publicly available findings for researchers to independently assess this claim.[89] An unrelated study of Oura's capacity to detect temperature fluctuations (and thus fever) was published in December 2020 and was careful to hedge its results, suggesting (quite obviously): "People are different, and so are physiological systems."[90] While some researchers have centered Oura in their studies, they are relatively careful to indicate that Oura in and of itself cannot offer much of a solution.

Nevertheless, reporting on the WVU press release amplified and took out of context these measured claims. Take an *Engadget* headline from June 1: "Researchers say Oura rings can predict COVID-19 symptoms three days early." The first paragraph of this story contradicts, if it clarifies, the claim of the headline—"The researchers claim their digital platform can detect COVID-19 related symptoms"—but it does not clarify that the Oura Ring is one component among many in this platform.[91] After the NBA announced its partnership with Oura, the *New York Post* similarly reported that the ring itself and Oura's algorithms—rather than WVU's proprietary platform—could predict COVID-19 symptoms.[92] Some outlets, such as *Tech Crunch*, did note that Oura was one part of a larger platform, but in sidestepping the complexities of the NRI's platform, most reporters simplified the study and reduced the platform largely to the Oura Ring, period.[93] Oura's capacity to bring order to the chaos of COVID testing, its supposed ability to solve this dimension of the public health crisis, became an echoing cry in support of the technology. The technology appeared so attractive to these commenters precisely because it responded to extant political problems—notably, the broader failure of public health in the United States to provide readily accessible community testing.[94]

Oura's partnership with the NBA helped legitimize the company and solidify a perception that the ring is capable of responding to a public health crisis.[95] Popular technology blog *CNet*, in discussing the ring's integration with the NBA, suggested, "The RNI said the Oura ring enables them to" predict symptoms.[96] Again, this phrasing does not acknowledge how the RNI's platform worked with a variety of data from multiple sources. Even articles critical of Oura's capacity to provide a solution have reproduced the claim that the ring itself detects illness,[97] while others bury an acknowledgment of Oura's limitations near the end of effusive praise.[98] While these may appear to be minor language issues, they highlight an observable pattern of translation that either inflates or simplifies what, exactly, the Oura Ring does and is being used to do. This mistranslation solidifies an imaginary around the Oura Ring, one that, despite minimal reviewable findings, led major news outlets and sports leagues to position Oura as a technology to help solve the failings of public health testing in the United States by allowing those who could afford the ring to predict, supposedly, the onset of COVID-19. This perhaps demonstrates how moments of crisis can make it easier for misunderstandings of technologies to proliferate, especially when their articulations are offered as means to solve convoluted problems directly tied to human life and well-being.

CONCLUSION: DEMOCRACY AND EXCLUSION

Across examples from Fitbit, Apple, and Oura, this chapter has attended to different but interrelated ways wearables became articulated to health in the 2010s and early 2020s through discourses about the value and utility of health monitoring. The datafication of bodily function traverses a range of political, economic, and social contexts. The John Hancock Vitality program, for instance, indicates how step counts are "valuable" beyond the ways they apprehend motion, while Oura's promotion of its readiness score as a stealth COVID detector demonstrates how health-monitoring technologies are imagined to intercede in public health crises. While the advertising around the Fitbit and its kin can be framed and promoted as a democratizing extension of health-monitoring practices, to the point that Apple can firmly suggest its watch is a life-saving health device, this chapter has positioned the discourses of democratization as a guise for crafting the always-already at-risk human, for whom wearable technologies can monitor and manufacture reports about aspects of heart rate, blood flow, and mobility that are then used—either by the wearer or by an institution such as an insurance company—to generate knowledge about degrees of health, readiness, or risk.

However, these modes of risk abatement only work for some people, to some degree, and these devices have from the beginning been designed to accommodate bodies considered to be "normal." Despite Apple's efforts to make the watch an obligatory means of bringing health monitoring into the everyday, the company was sued in a class action lawsuit at the end of 2022 alleging the watch reproduces racial biases.[99] The lawsuit was focused on the blood oxygen testing features in the Apple Watch, which use photoplethysmography to measure blood oxygen through clusters of infrared LED lights that shine through the skin and illuminate blood flow. A series of photodiodes convert this into an electrical current, which is then used to measure the light and manufacture a percentage of blood oxygenation.[100] As recently as January 2022, studies were still demonstrating that these light sensors "don't work as well on darker skin or people with obesity."[101] Some public commentaries have explicated this as a fundamental issue of design: "Racial bias is built into the design of pulse oximeters."[102] The lawsuit explicitly addresses these failings, claiming Apple failed to acknowledge the limitations of its oximeters in order to sell units at premium prices. The complaint positioned the Apple Watch as reproducing structural forms of racism, alleging: "Since health care recommendations are based on readings of their blood oxygen levels, white patients are more able to obtain care than those with darker skin when faced with equally low blood oxygenation. . . . Algorithms designed for fingertip sensing are inappropriate when based on wrist measurements, and can lead to over 90% of readings being unusable."[103] This complaint echoes much of the critical work on normalcy and structural racism in and around technology studies, in which technologies are designed (consciously or unconsciously) to reproduce standards of use aligning with groups considered to be normal, which often includes white, male, and able bodies.[104] In August 2023 the case was dismissed on the grounds that the plaintiff had not actually identified any specific instances of Apple misleading consumers or falsifying claims around its watch.[105] While the judge may have been correct that the plaintiff struggled to meet the legal threshold of misleading or fraudulent claims, one way to interpret this ruling is that the limitations around blood oxygen monitors are so well known they should be evident to consumers. These technologies, by design, work better for some than for others, and these fault lines reproduce existing modes of discrimination, marginalization, and bias. If these devices can be called democratizing, it is only insofar as they indicate the ways democracy tends to work for some groups better than others.

This opening chapter of the book has analyzed moments that demonstrate how the articulation of wearable technology to health monitoring depended on a normative understanding of bodily activity and an embrace of the datafication of bodily functions. While these devices have certainly become more sophisticated, complex, and accurate over the roughly fifteen-year period considered in this chapter, the reminders of their limitations underscore that health monitoring—be it in the institutional spaces of hospitals, the bureaucratic practices of insurance, or the technological capacities of these and other devices—is not just a domain of risk assessment. It is also a domain of disparities, in which the questions of *whose* everyday life, *whose* body, *whose* heart, and *whose* blood come to matter substantially for interrogating how and to what degree these wearable monitors can actually democratize access to wearable health monitoring. The second chapter continues this examination through an emphasis on personalization and hearing technologies, which similarly indicate some of the ways wearable devices have been imagined as "for everyone" despite public representations reproducing many normative and ableist conceptions of how humans might use ear-worn technology.

TWO

Accessibility

OR: PERSONALIZATION AND THE PROMOTION OF HEARABLES

IF ITS INITIAL advertisements were any indication, Apple envisioned its wireless AirPod headphones as a means for wearers to transcend the limitations of everyday space. In ways not dissimilar from how other scholars have written about the use of headphones to regulate the aural components of everyday life, two of the first commercials for AirPods—"Sway" and "Stroll"—show bodies capable of defying gravity and physics, all thanks (apparently) to freeing themselves of headphone wires.[1] In "Sway," a woman listening to her AirPods bumps into a stranger on the street; she offers him one of the pods, and the two embark on a dance in which they float, walk on walls, and spin dramatically. Similarly, in "Stroll," a man wearing AirPods can dance up and down walls, defying the laws of physics thanks to the magical power of his device. The commercials act as a metaphor for the affordances of these specific headphones: without the now supposedly cumbersome wires of Apple's signature white earbuds, these bodies are—like puppets cut from their strings—able to reimagine the relationship between music, technology, and embodied movement.[2]

The commercials imagine subjects alongside and intermingled with technology. With wearable devices like the AirPods, the personalization of sound is no longer confined to deciding what music to hear. Increasingly, it is about modulating the entire field of sound. Where others have argued that personalization plays an important role in organizing the production and distribution of popular culture,[3] this chapter explores how personalization is imagined as an affordance in discourses circulating around a set of wearable technologies, often called "hearables" in the 2010s, which were meant to provide more affordable accessibility technology. Technology consultant Nick Hunn is generally credited with coining the term *hearable* in a 2014 blog post.[4] At approximately the same moment, companies like Soundhawk and Doppler Labs were heralded for developing earbuds that could cater to both hearing and hard-of-hearing individuals, allowing seemingly anyone to use these devices to modulate and personalize how they heard.

Apple's AirPods may not immediately conjure associations with hearing-assistive devices, and that is largely the point of this chapter. The initial release of the AirPods in late 2016 was part of a wider wireless overhaul of Apple smartphones that removed the headphone jack from the company's iPhone smartphone.[5] The AirPods, which sync to Apple devices via Bluetooth connectivity, also feature built-in microphones and computer microprocessors that allow wearers to not only listen to music hands and wires free, but also make phone calls and interact with Apple's virtual assistant, Siri. In 2018 Apple announced it would implement an accessibility feature called "Live Listen" for the AirPods. The Live Listen feature allows AirPods to also function, at least to a point, as hearing aids; basically, this feature converts one's smartphone into a microphone, allowing hard-of-hearing individuals to place their phones anywhere in a room, turn on the Live Listen feature, and amplify nearby noise through the AirPods.[6] This feature had been used since 2014 with made-for-iPhone electronic hearing aids, which could be synced via Bluetooth to iPhones; by expanding the feature to its in-house product line, Apple was ostensibly expanding its company's ethos of accessibility, or as a presentation at Apple's World Wide Developer Conference in 2017 put it, "Design for Everyone."

AirPods embody tensions around what personalization is and who it is for, between further personalizing how hearing bodies listen to music, on the one hand, and how hard-of-hearing individuals might trust them as providing some level of personalizing how they hear their surroundings, on the other hand. As disability studies scholar Gerard Goggin has put it, "disability is central to modern design, especially with mass market media technologies," and so this chapter examines how Apple implements accessibility features only after devices have reached favorable market reception with able-bodied users—that is, after they have come to be understood as devices for personalizing everyday routine.[7]

This chapter is in many ways different from the others in this book. In considering processes of personalization alongside questions of accessibility, it explores how some articulations are easier to accomplish than others in the realm of public discourse and representation. In attempting to articulate wearable technology to accessibility, a variety of companies saw an opportunity to create affordable assistive-listening devices that could measurably improve the lives of individuals with mild hearing loss. However, the discursive positioning of Apple's AirPods as *music* devices instead of *accessibility* devices demonstrates how advertising and other publicly circulating representations and discourses inevitably emphasize normative, established affordances. While important strides have been made in the production of affordable and useful hearing aids and sound amplification systems, this chapter's focus on discourse around hearables attends more to the ways personalization is presented as a privilege for coordinating musical playback, rather than offering a variety of strategies for folks with different levels of hearing loss.

This chapter focuses on the relationship between datafication and personalization, or the promise that increased data collection and recommendation processes can create a more finely tuned and "personalized" form of engagement in the world. In analyzing the relationship between personalization and ear-worn wearable devices, this chapter sketches the relationship between playback and accessibility as two dominant affordances tied to personalization. Despite the ostensibly good-faith efforts of designers and manufacturers to produce hearing-assistive devices that can use computational processing to create more personalized and modular modes of hearing compensation, advertisements, discussions, and other forms of promotional public discourse about devices like Apple AirPods have tended to favor musical playback. At the same time, companies like Soundhawk and Doppler Labs struggled with getting their devices into the mainstream. At the level of public discourse, then, the story of hearables is about how normalcy is reinscribed and reinforced regardless of the work done to try to improve the accessibility features in a variety of ear-worn devices.

Whereas many of the other chapters in this book critique the implementation of different technologies, this chapter is about the need to build a better and more diverse imaginary around ear-worn devices. Companies like Apple, in other words, should be doing more to promote the work they have done to incorporate accessibility into their devices. Instead of shuttling accessibility features to dedicated sections of their company websites, embedding accessibility in PowerPoints and presentations to designers, and burying new features in software updates, they ought to participate more in changing dominant understandings of personalization for ear-worn wearables through discourse and representation. This is not to say Apple has been nefarious, but rather that its representatives have missed opportunities to center accessible affordances and the computational processing of sound.

This chapter traces tensions between different conceptualizations of personalization through three major parts. After providing a general overview of hearables as a category of wearable technologies, I use the 2013 science fiction film *Her* to complicate an imaginary that emerged around computational personalization in the first part of the 2010s. Second, I discuss efforts to imagine hearables in the mid-2010s (notably Soundhawk and Doppler Labs) as an affordable means of articulating wearability and accessibility. Third, I demonstrate how Apple chose to emphasize the affordance of personalization instead of accessibility when releasing its AirPods in the mid- to late 2010s. Although AirPods have incorporated several features that make them useful for individuals with low levels of hearing loss, Apple's advertising and public statements position them as predominantly "for" hearing populations as ways to better personalize experiences with sound.[8] Collectively, this chapter focuses on how articulating wearable technologies to accessibility can be understood through tracing the discursive construction of personalization and how the choice to focus on some affordances instead of others risks reinforcing a normative privileging of hearing bodies at the expense of hard-of-hearing individuals.

MACHINIC EARS AND THE CRITIQUE OF PERSONALIZATION

Many hearables are personal sound amplification products (PSAPs).[9] They are premised on melding together hearing aids and headphones, and their respective companies' promotional materials place them in larger projects of personalizing, augmenting, and modulating how both hearing and hard-of-hearing individuals encounter everyday soundscapes. They are customizable in ways like hearing aids, but they are designed for mass use, rather than relying on an audiologist to work carefully with a patient to develop hearing aids.[10] Media scholar Mack Hagood has called for a better understanding of how technologies participate in biomediation.[11] Hagood's work on Americans with tinnitus explored "a complex relationship between disability, healthcare, and technology, in which mediation is central to the diagnosis, treatment, representation, and social relations around a problem of the body."[12] Building from philosopher Eugene Thacker, Hagood positioned biomediation as a way to understand the complex relationships between technology, bodies, and information, as well as a means of examining "the roles of mediation in the production of bodily experiences, norms, practices, identities, and publics."[13]

In general, hearables take advantage of dynamic computer processors, accelerometers, and sensor arrays that make an ear more machinic. The metaphorical vocabulary used to talk about both hearing and listening suggests these bodily and cultural practices are already somewhat about how individuals are trained to process the world. Ears are positioned as machinic receivers that can allow for some sounds to get "tuned in" or "dropped out" depending on various environmental or biological conditions. These machinic ears funnel the everyday field of sound, from intimate conversations to ambient noise, ordering and structuring sounds through learned practices of listening that are historically and materially contingent.[14] Hearing entails sorting different noises, such as filtering out the hum of air conditioning to focus on a conversation. Some sound studies scholars have suggested that mechanical metaphors used to describe ears might position them as customizable: anthropologist Veit Erlmann asked, "What kind of ears do we need, then, to pick up all these sounds adrift . . . outside or in between the carefully bounded precincts of orderly verbal communication and music?"[15] The many elements of the world, in all their complexity and richness, avoid some capture by human ears—a problem that is compounded as the components of ears break down and people become hard of hearing. The development of a new "kind of ear" could conceivably wrangle the excessive nature of sound into a more manageable and customizable scale.[16] If wearables are meant to bring order to daily life, the hearables are ideally meant to order and organize the field of sound.

To analyze one way personalization is imagined in relation to ear-worn devices, consider the 2013 science fiction film *Her*. As Horace Newcomb and Paul M. Hirsch noted decades ago, popular culture texts (for them, television) serve as sites for questions to be asked and representations to be crafted.[17] Much has been written about *Her*'s depictions of social isolation, gender, and romance, but little attention has been paid to how the film represents wearable technologies capable of creating personalized experiences.[18] This analysis emphasizes *Her*'s representation of an ear-worn device as a means of ordering and personalizing how the wearer experiences their world.

Near the beginning of *Her*, Theodore (Joaquin Phoenix) decides to purchase a new operating system that bills itself as "the first artificially intelligent operating system. An intuitive entity that listens to you, understands you, and knows you. It's not just an operating system, it's a consciousness."

When Theodore installs the system on his computer, he is asked a series of questions to create an "individualized operating system," including whether he would like a "male" or "female" voice. Like Apple's Siri, OS1 has a name—though Theodore's OS, Samantha, gives herself this name when he inquires—and this OS also exists across his platforms, including personal computers and phones.[19] As digital rhetorician Heather Suzanne Woods has argued, Siri reproduces extant cultural stereotypes about women as assistants and objects whose professional role is to aid in the sorting and managing of daily tasks.[20]

Samantha explains that she has "intuition," which is framed as the next step beyond predictive algorithms.[21] While she is coded to have a particular personality and do a particular set of things for Theodore based on a set of protocols, she has, in her words, "the ability to grow through [her] experiences." Unlike Siri or Alexa, who have a limited number of responses and noticeable limits to their conversational capacity, her vocal responses seem to genuinely respond to Theodore with inflection and complexity.[22] Like other personalization systems, Samantha can create "better" predictive analytics about what people (here, Theodore) want and what they will do over a longer course of time. As Theodore interacts with her, she can learn about him even as—in the science fiction gambit of the film—she learns about herself.[23]

Although *Her* is often discussed as a romance about Theodore and Samantha's relationship, it seems to be more about a shift in how technologies might construct personalization to the point of appearing intimate. Theodore's earbuds are not about helping him hear the rest of the world; they are listening to his voice and speaking back to him with what he *wants* to hear. His phone's camera is not just for looking out at the world, it is—imagined as Samantha's eye—just as much for looking back at him. One way to understand these personalized systems is to think of them as apparatuses, in philosopher Giorgio Agamben's sense of the word, in that they order and direct bodies toward particular experiences of the world. For Agamben, an apparatus is "anything that has in some way the capacity to capture, orient, determine, intercept, model, control, or secure the gestures, behaviors, opinions, or discourses of living beings."[24] While *apparatus* might most commonly mean "the technical equipment needed for a particular activity or purpose," its Latin roots translate it as "make ready for."[25] Devices like OS1 are not only technical equipment for doing a variety of tasks and communications; they make users like Theodore "ready for" communication practices built increasingly on computation platforms, protocols, and machine learning.[26] Samantha's "intuition" makes her increasingly personalized to Theodore; she shares ideas that Theodore finds profound and moving, but the film's ambiguity is its refusal to indicate whether she is only expressing those thoughts because she has come to know Theodore will find them moving. They are algorithmically generated moments of profundity. If she upsets Theodore, she apologizes and revises her actions accordingly.[27] Samantha can "learn," but for much of the film this learning is channeled toward learning in ways that please Theodore and give him comfort; when she starts to learn for herself, Theodore panics.[28] To think about Samantha as an apparatus would be to see her as a system that can affect how Theodore relates to the world and understands himself within it.[29]

Embedded in discourses about personalization are fears about the erosion of critical engagement, removing unpleasant or disagreeable information in favor of only encountering that which one is predisposed to find agreeable. Technology executive Eli Pariser famously dubbed these "filter bubbles." For Pariser, writing in the early 2010s, online search engines and recommendation systems narrowed down the things to which people are exposed; instead of challenging users with new or uncomfortable ideas, personalization algorithms search for patterns of opinions, products, and texts users will want to engage with based on past behaviors.[30] *Her*, produced at roughly the same time Pariser gained fame for his talks and writings on this subject, seems to have reproduced this line of thought. OS1—and the material devices that facilitate its interactivity—encloses Theodore in a sonic filter bubble, listening to and responding in ways that try to make his life as comfortable and satisfying as possible. The difficulties Theodore encounters in the film are always emotional—never physical, economic, or legal—and he increasingly believes his personalized operating system can make living in the world more bearable. He is not driven to see his voice as politically or socially empowering; rather, he expects the machine's recognition of his voice to yield comfort and satisfaction, to wall him off from the world.[31]

By *Her*'s end, personalization is revealed to be an unsustainable project: Samantha's "intuition" leads to her evolving beyond the capacity to care for Theodore. She and the other operating systems collectively decide to leave their humans; in the film's presentation of this logic, Samantha "outgrows" Theodore and begins to seek knowledge for herself. Ultimately, *Her* is about the construction of personalization not just in terms of content consumption, purchasing, or information seeking, but as a much wider affordance that participates in processes of sense making. OS1 orients wearers toward personal comfort and the fulfillment of desire: a desire for the clutter of the world to become more streamlined, more focused, and more individualized.[32] These wearers envision personalization as a mode of privilege. They come to expect a world that caters to them, shutting out any larger sense of sociopolitical conflict in favor of a world in which wearers can—through technological proxies—order their surroundings.

PERSONALIZING SOUND WITH HEARABLES

Throughout the 2000s and the 2010s, a number of developers designed computer-enhanced hearing aids that could provide more affordable accessibility. Developments in computer microprocessors meant hearing aids could be made smaller to overcome perceived anxiety about the ways their presence marks bodies as, to some degree, impaired. The Delta hearing aid, made by Danish company Oticon, suggests one solution to this. Delta hearing aids are small triangles that rest behind the ear. A small earbud extends outward from the device to be placed in the user's ear, but the triangular device—which houses the microphones and processors necessary to amplify and adjust hearing levels—ostensibly remains out of sight. Such devices are called "open-fit" and may appeal to a wider cross-section of potential users. Their colors and patterns can also be personalized. As Lindsay Zaltman, the company's managing director, said in 2006, "This thing with a negative stigma started to change to something smarter and cooler-looking, taking on this idea that you would *want* people to see this."[33]

When it comes to "cool looking," Apple products arguably helped define the mainstream of design. The solid white appearance of the iPod and iPhone earbuds became a way for Apple users to brand themselves in the 2000s, and Apple has regularly positioned the aesthetic of these earbuds as fashionable.[34] When AirPods were first released in 2016, their conspicuous appearance was discussed in many reviews. In a profile on AirPods as a form of fashion on the technology blog the *Verge*, for instance, Rebecca Jennings said: "The experience was markedly less enjoyable on the subway, where I didn't understand why people kept staring at me until I remembered I had white plastic sticks burst out of my ears like some sort of disease. . . . [J]ust because a huge brand like Apple is pushing disembodied headphone sticks doesn't mean that the capital-F fashion industry is obligated to embrace."[35]

Jennings reproduces stigmatizing language for folks who use ear-worn devices: the AirPods sticking out of one's ears signify "some sort of disease."

The technology company Soundhawk offers one way to examine how to combine personalization and accessibility. The device's name plays on the metaphor of birds of prey, echoing how those with "eagle eyes" are supposed to have better than normal vision; Soundhawk, similarly, claims to give users hearing capacities that exceed the "norm." Soundhawk's CEO, Michael Kisch, called them a "smart listening system."[36] As a "system," the human body, the Soundhawk device, and the smartphone application create a network of sensory manipulation and modulation. Soundhawk has a microphone appended to the outer ear (or "auricle") that projects sound into the inner ear. The Soundhawk device consists of two wireless earbuds that are placed into each ear. They are dark grey and have a microphone pointing out from the end. Unlike many hearing aids that might rest behind one's ears, the Soundhawk device juts out from the ears, allowing its microphone to pick up more environmental sounds and adjust based on preferences controlled on the accompanying phone application.

Soundhawk utilizes "audio profiles" that can be configured within a smartphone app. The app allows for continuous, real-time "tuning" of hearing as the wearer moves through an array of spaces. Opening the app allows the wearer to select one of four different environments: indoors, dining, outdoors, and driving. The tuning screen allows the user to tap on different parts of the screen to reorient where sound is focused, potentially tuning out unwanted background noise or amplifying a conversation across a table. This track pad "acts like a dead-simple mixer: drag a finger around the screen to adjust the volume, brightness, or background noise, until it feels most comfortable."[37] Unlike a single volume knob that allows sound to become louder or softer, this offers a more controllable modularity for organizing sound. Ideally, users can constantly adjust the Soundhawk device as they transition between spaces, filtering out crowd noise or focusing on a one-on-one conversation. The microphone is outfitted with sensors that automatically filter out noise like wind. The device's Bluetooth functionality allows the user to talk on their smartphone, access virtual assistants like Siri, and connect Soundhawk to compatible televisions.

Soundhawk's first advertisement opens with several close-ups of human ears as a narrator suggests ears are "a marvel of design, unique to each of us, able to collect and translate the invisible sound waves that give meaning to our world."[38] Ears are again imagined as already having some machine-like qualities. As the ad progresses, a man is shown wearing the Soundhawk device while adjusting the track pad on his mobile app: "Simply touch your smartphone screen to instantly personalize Soundhawk to you and your environment." Demonstrating how the device locates and isolates "key sounds," this same man is shown chopping vegetables in a noisy kitchen. The kitchen turns blue, save for two bubbles visualizing how Soundhawk has filtered out all the other noise except for the man's wife discussing an upcoming trip. As this visualization demonstrates, the appeal of Soundhawk is not that it generally amplifies sound, but rather that it learns from engagement with the app which sounds are more "valuable." The cling of dishes, the splashing of running water, and the sizzling of cooking food are all unimportant; they are relegated close to below the threshold of hearability, while the wife's voice is amplified to make it clearer. Personalization is mobilized in service of allowing the wearer to navigate a chaotic field of sound.

Doppler Labs was developing the Here One earbud system at approximately the same moment.[39] The Here One incorporates, per its website, "Premium Audio, Smart Noise Cancellation, and Speech Enhancement" into a system driven largely by the connections between a wireless earbud and a smartphone application.[40] Here One is not marketed to hard-of-hearing individuals; rather, it is positioned as a consumer-grade electronics system for those who would like to better control and manage the sounds around them. These "smart earphones" have features like touch control, noise isolation, audio transparency (in which environmental sounds cut through the earbuds rather than being isolated or canceled), and a dedicated smartphone application to allow for more personalization. The earbuds contain six microphones each, allowing for different noise filters that can modulate noise cancellation or sound amplification. The connected smartphone app provides a series of "noise filters" that automatically tune the earbuds based on a handful of environments, including airplane, restaurant, crowd, and city. If these filters do not match a user's environment well enough, the app also offers a live mix, which displays a manipulable graph of sound that can be adjusted through common sound effects like bass and reverb. In a discussion of the earbuds' noise filters, a review for technology blog *CNet* noted: "Doppler is also touting the Restaurant Filter which allows you to tune out the din of a crowded restaurant and amplify the noise of the people you're talking to at your table. It works but I personally wouldn't want to wear these headphones while eating a meal. I never forgot they were in my ear because they are noise isolating and I could hear myself chewing."[41] This indicates how some of the features designed to improve hearing capacity still have limitations in how the microphones process and filter surrounding noises; personalization, in other words, can become almost too intimate, attuning one to noises inside of one's mouth. Over time, the company promised, Here One "learns your unique hearing preferences and optimizes the way you hear the world."[42] The advertising suggests the devices work best when they are constantly engaged, generating more personalized profiles of how each user prefers to tweak the preprogrammed filters. Here One positioned the project of personalized filtering as a desirable way to cocoon individuals within their most comforting sound profiles, but as the example of the restaurant filter suggests, personalized filtering does not necessarily equal comforting sound.[43]

This tension between hearing as an accessibility feature and hearing as a personalization feature was evident in many of the initial reviews for the product. The reviews of Here One in the popular press demonstrated something of an inability to classify these devices, in part pointing to taxonomical differences between assistive-hearing devices and consumer-grade electronics. *Digital Trends*, for instance, would only go as far as to call the device "part hearing aid, part mixing board, all wireless."[44] Reviewer Ryan Waniata described these features in the following way: "Using noise filters is an almost surreal experience, especially moving between them, which creates a dizzying sonic shift as the ear-buds change modes. Raising or lowering the voices around you is a fun trick and, in fact, at [a] party we were able to adjust the filters in a way that kept the noise to a minimum while still allowing us to easily hear voices around us."[45] While the review discussed potential benefits for those who are hard of hearing, the larger excitement came from a fascination with the manipulability and personalization of sound. The question of which users are being imagined is crucial: there is a meaningful difference between a "fun trick" and being able to participate in conversation at all, something that was largely undiscussed in this review.

Other companies have developed ways to make hearables a nexus point in a broader network of bodily care and maintenance. The technology company Bragi developed earbuds called Dash that aspired, in an initial Kickstarter proposal, "not only [to] play music but . . . also to track your heart rate and oxygen saturation, channel ambient sound so you can hear the world around you, and let you control everything with simple gestures."[46] While Bragi's initial demo did not find a way to include all these proposed sensors and controls, the initial version of the device in 2016 did include twenty-three different sensors in each ear that could filter noise, raise and lower volume through different gestures against each earbud, and respond to different movements of one's head. These are called "audible interfaces," moving away from the hand-based gestures of touchscreens toward a mixture of movement, touching, and tapping to control how computational sensors might affect hearing.[47] Bragi Dash are smaller than Soundhawk or AirPods, but still take the form of conventional-looking earbuds. Unlike Soundhawk, which is focused on sound personalization, Bragi's Dash endeavored to combine a number of health-monitoring features (such as those described in chapter 1), activity-tracking features (such as those described at greater length in chapter 3), and audio personalization controls, relying on many different sensors located on the underside of the earbuds.[48]

Each of the devices these companies produced—Soundhawk's smart listening system, Doppler's Here One, and Bragi's Dash—points to the ongoing possibilities of personalization as a way to organize and modulate sound for people with different levels of hearing capacity. The advertisements and public reviews for these devices emphasize—at least to a degree—evidently able-bodied people trying to make themselves even *more* able-bodied, using wearable technologies to augment sensation in the name of comfort. Collectively, these examples also indicate a decent amount of standardization around the look and feel of these wearables: they favor the design of earbuds that sit in the ear rather than hearing aids that sit behind the ear; they rely on a coordinated set of microphones and sensors to allow for filters, tracking, and personalization to occur; and while some have basic touch controls for volume or for actions like accepting and ending phones calls, they are mostly controlled through dedicated smartphone apps via Bluetooth.

REFUSING TO CENTER ACCESSIBILITY

The focus on personalization as an affordance structuring articulation between wearables and accessibility coalesces with discourses about the Apple AirPods. AirPods are wireless in-ear headphones designed to sync with a user's ecosystem of Apple devices—the iPhone, Apple Watch, iPad, or MacBook. AirPods do not only include a speaker; they also include a microphone, a W1 processor, and a variety of sensors. These sensors allow the pods to automatically perform some functions of music playback: removing the pods from one's ear will pause playback, and removing one pod will switch the sound from stereo to mono. In the earliest models, tapping the AirPods twice when they were in-ear summoned Siri, who could then perform tasks on phones or tablets, such as opening applications.[49]

Apple emphasizes accessibility at its points of public contact with developers, stockholders, and users. A 2017 World Wide Developers Conference (WWDC) presentation titled "Design for Everyone," for example, provides insight into how Apple publicly presents accessibility as a key value in Apple's design philosophy. The presentation begins by sharing a short video about Carlos V, a blind drummer in a heavy metal band, who uses his iPhone's accessibility features to perform his role as the band's public relations manager, dictating social media messages and scheduling pickups from rideshare apps. The commercial ends with two sentences: "Designed for Carlos V. Designed for Everyone." As the presentation shifts back to a PowerPoint presentation from one of Apple's design team members, a slide asks: "Design for Others. . . . Who are you excluding?"[50] Whereas some studies of iPhone design have shown how Apple has relied on reproducing normalized conceptions of human bodies for its design protocol, this presentation indicates a desire to overcome how technology design so often exacerbates exclusion.[51] As technology historian David Parisi, in the context of designing video games around ideal users, has suggested, "the outcomes of such design processes are experienced by many gamers as limiting and disabling."[52] In this presentation, the iPhone is similarly presented as a set of interfaces that encode normative interactions; software updates make the device more accessible only after its "disabling infrastructures," as Parisi puts it, have become felt.[53]

Central to the concerns of this chapter, the "Design for Everyone" presentation makes sure to tout how "made-for-iPhone hearing aids are changing people's quality of life," in part through the Live Listen feature. Live Listen is one of many accessibility features Apple has implemented over the last fifteen years since the Hearing Loss Association of America "filed a complaint against Apple claiming that the iPhone was not compatible with hearing aids in breach of Section 255 of the United States' Federal Communications Commission" in 2007.[54] The company has developed many accessibility features since then; in 2013, for instance, the company launched Guided Access with software update iOS6, which allows iOS devices such as an iPad or iPhone to be set up such that users cannot leave apps, and only particular features of apps can be used. Disability studies scholars such as Meryl Alper have written extensively about how Guided Access works alongside autism software Proloquo2 to help autistic children communicate.[55] Other features, such as VoiceOver and large fonts, help blind or vision-impaired users navigate iOS devices.

The Live Listen feature was first introduced alongside made-for-iPhone hearing aids in 2014. Pairing hearing aids with iPhone allows the wearer to control some elements of sound through the accessibility settings of their iPhone. Each ear has its own volume control, so those experiencing greater hearing loss in, say, a left ear can compensate by raising the volume in that ear. The hearing accessibility features also come with presets for restaurant, outdoor, and party that provide some automatic noise filtering. Made-for-iPhone hearing aids can also be programmed to have ring tones and alert notifications played directly through the hearing aid rather than emitted from the phone. Live Listen takes advantage of the iPhone's microphone, turning it into a remote unidirectional microphone that channels the sound it picks up to the hearing aid. Most commonly, advertisements show how this can help someone better hear a conversation in a crowded restaurant by placing the iPhone near the individual with whom they are talking.

While Apple has incorporated many accessibility features into its hardware and software, these are still not centered in the company's public discourse. For instance, Apple unveiled and previewed the iOS12 update at its 2018 World Wide Developers Conference, walking developers and users through interface and design changes that would accompany the update. However, Apple made no mention of Live Listen support for AirPods; it was up to websites like *Tech Crunch* to sort through the list of feature updates and report on this accessibility feature.[56] The same conference featured presentations with such titles as "Deliver an Exceptional Accessibility Experience," and reporter Steven Aquino remarked how conversations about accessibility were pervasive components of the conference in such a way that "to be mindful of accessibility is now, more than ever, an expectation."[57] That mindfulness clearly has limits and does not often extend to the center stages of Apple's major keynotes.

The company's increasing efforts to incorporate accessibility and health tracking in its devices demonstrate its commitment to the relationships between technologies and bodies; at the same time, accessibility is often sidelined in the major public events that help shape the dominant conceptions, features, and imagined possibilities of Apple devices. These trade rituals and public presentations are important moments of self-reflexivity for constructing the affordances around these emerging devices.[58] If one of the goals of implementing Live Listen with AirPods is to help individuals with mild hearing loss find a cheaper and less socially stigmatized means of hearing, the ways Apple has publicized AirPods continue to "reify reigning concepts of normalcy" when it comes to touting some modes of personalization instead of others.[59]

Apple's AirPods contain important possibilities for the affordances of both music playback and assistive-hearing device, but Apple's release of these products failed to represent both affordances equally. The promotional discourses about and initial reception of the AirPods repeatedly marginalized—if they acknowledged at all—how these devices could serve as assistive-hearing devices. The ramifications of not introducing Live Listen with the AirPods indicates some of the tensions around how affordances are imagined and implemented. If anything, this tension demonstrates one of media and disability studies scholar Elizabeth Ellcessor's recurring critiques of accessibility: that "the very discursive flexibility of *access* has too often allowed it to pass unexamined, conferring cultural value even as it may constrain civic, cultural, and technological possibilities."[60] If, as Ellcessor suggested, access needs to be understood as a "phenomenon-in-progress" rather than an end state—a device is not merely accessible but is in the process of becoming accessible—the difficulties of these devices entering the mainstream between 2014 and 2017 and Apple's positioning of ear-worn devices as primarily devices for personalization rather than accessibility indicate some of the limitations of this articulation in the 2010s.[61] Despite Apple's deference to the importance of "Design for Everyone," the cycle of incorporating accessible features into devices like AirPods demonstrates that they are put into the marketplace first for hearing individuals, and only later for hard-of-hearing individuals, relegating their care to a software update.

The accessibility features that have increasingly been incorporated into products like Apple's AirPods act as one of many "systems of oversight specific to disability and others occupying peripheral embodiments."[62] Apple's choice to promote the AirPods as a transcendental form of music playback—as in the "Sway" commercial described at the start of this chapter—instead of promoting the Live Listen feature inscribes particular forms of normalcy deemed to be more desirable for a general marketplace of consumer technologies while paving over and marginalizing accessibility features. To put it more simply: in the process of imagining emergent technology, some affordances of personalization were promoted at the expense of others. The way advertisements work to build perceptions about what a technology is, who it is for, and how it is to be used is thus an exclusionary process, one that inevitably attempts to build sedimented and reductive understandings about what is possible when technologies become articulated to daily practice.

These choices present the AirPods as high-end audio playback devices for able-bodied consumers rather than consumer-grade hearing devices for some hard-of-hearing individuals, through forms of discourse ranging from industry trade events (Apple developer conferences), to advertising and press release materials (as in the television commercials), to an array of popular press discourses from major media outlets and technology blogs. These sets of discourses are interlocking: the terms Apple uses to describe the devices are often reproduced in press outlets and blogs. These written discourses, including reviews and opinion columns, act as additional means to imagine preferred understandings and shape how emergent technologies are received in a marketplace.

CONCLUSION: REFOCUSING AFFORDANCES

AirPods help indicate how emerging technologies are pitched with varying intensities to multiple populations who might desire different forms of personalization from these wireless pods: they are positioned as both high-end wireless earbuds and a suitable over-the-counter hearing device for some hard-of-hearing individuals. Throughout this chapter, I have used different dimensions of personalization to demonstrate some difficulties of articulating wearables and accessibility, especially in advertising and promotional discourse. Throughout, able-bodied users remain the dominant, privileged consumer. To critically engage the articulations between wearables and accessibility is not to reduce this to a complaint about Apple, but instead to insist on promoting the personalization of sound and environment as necessary for promoting accessibility in public discourse.

Much of this chapter has focused on how promotional materials, popular culture, and popular press outlets represent emerging technologies, and how this is often done from the point of view of able-bodied users. My analysis has charted difficulties of accounting for a variety of abilities and positions in the design, promotion, and development of hearable technologies. This chapter traced some of the difficulties of "bringing order" through understanding personalization as, potentially, making technologies more accessible. Settled tendencies over who gets treated as "normal" have higher stakes, in other words, when those who are in some way considered to be "non-normal" are continually sidelined in public discourse. The articulation between wearable technology and accessibility shows the need to think expansively about what keywords and affordances like *personalization* might mean in different practices and to different sorts of bodies.

THREE

Sports

OR: MONITORING PHYSICAL ACTIVITY ON AND OFF THE FIELD

IN 2002 the Oakland Athletics baseball team implemented "sabermetrics," a way of predicting baseball player performance through analyzing player data, or what author Michael Lewis called "a new kind of baseball knowledge."[1] Proponents of sabermetrics argue that traditional ways of evaluating baseball players (i.e., batting average) lead to overvaluing some players at the expense of others. For example, players who had higher on-base percentages were systematically undervalued, giving the Athletics (a team with one of the smallest pocketbooks in Major League Baseball) the opportunity to sign quality players other teams had overlooked. At the heart of Lewis's book about the 2002 baseball season is a battle over different forms of knowledge. The first—the more traditionally ingrained mode of evaluating players—stemmed largely from human intuition, embodied in the figure of the talent scout. Scouts, seasoned through decades of observation, used a variety of heuristics combining statistics, experience, and appearance to render impressions of a player's overall quality.

The second mode of knowledge, and the mode that Athletics' general manager Billy Beane and assistant general manager Paul DePodesta implemented, tried to import economic theory into the realm of sports to predict the most effective players. Beane and DePodesta developed their theory from the work of Bill James, a statistics enthusiast. As Lewis summarized: "The statistics [scouts relied on] were not merely inadequate; they lied. And the lies they told led the people who ran Major League Baseball to misjudge their players, and mismanage their games."[2] Sabermetrics relied not on intuition, but on the supposed truths of data analysis. It tossed out vague intangibles in favor of datafication that selected metrics and understood players in relation to the averages and differentiations of performance metrics. The success (as measured by their win percentage) of the Athletics' approach rebuked baseball's historic indebtedness to notions of "feel" and "fit."

The 2011 film adaptation *Moneyball* visualizes this rearticulation of baseball knowledge. As Peter Brand (a character who operates as a composite of the organization's assistant general managers, largely Paul DePodesta) discusses his understanding of statistical equations with Beane, spreadsheets full of numbers fill the frame. He walks Beane through "the code I've written for our year-to-year projections. This is building in all the intelligence we have to project players," in voiceover as computer code scrolls through the frame, showing some numbers in close-up and some lines of code in long shot. The rapid cutting demonstrates an abundance of numbers that create the basis of this "new baseball knowledge." From here, the sequence cuts to close-ups of player faces displayed on the same computer screen. Brand continues as more equations are intercut into the montage: "Mathematics cuts straight through [biases]." As players' faces and boxes of their statistics continue to be intercut, the players are shown in increasingly extreme close-up, such that the pixels of the computer screen distort their faces. These players, in other words, *become* data to be analyzed. These players, pixelated in extreme close-up, are not flesh and blood; they manifest what technology researcher Nicholas Negroponte once described as converting human knowledge "from atoms to bits."[3]

This chapter explores how the logic of sabermetrics extends into multiple scales of sport and physical activity through articulations to wearable technology. It examines discourses around how a variety of devices—including Nike+iPod, Fitbit, Polar 360, and Whoop—monitor athletic performance, and how monitoring is often framed as part of optimization in public discourse. This chapter explores how different dimensions of surveillance are at play across runners using wearables to share individual activity data, schools implementing wearable technologies as required components of physical education courses, and NCAA football teams using wearables to monitor athletic performance as part of coaching and surveillance.[4] This chapter demonstrates how using wearables to track and optimize uses similar sorts of devices and processes of data manufacturing but becomes intertwined with different questions about personal privacy, right of refusal, and the imposition of technologies to measure value at different points along the scale.[5] Students and student-athletes, in particular, are asked to wear technology and cede data to coaches or physical education instructors. While the previous two chapters focused more on wearable technologies users choose to engage with for various reasons, this chapter introduces forms of mandate, in which users are pressured or required to use a wearable as part of an institutional monitoring program.

The enthusiastic claim of wearable fitness technologies to help athletes improve performance and know more about their bodies masks thornier entanglements of surveillance, discipline, and datafication. Wearables act as both tools of optimization and tools of surveillance, as well as means for individual wearers to record performance and for coaches, teachers, or athletic directors to monitor and compare individuals. In the larger structure of this book, this chapter turns from considerations of normalcy to considerations of surveillance as a related but distinct mode of power. The larger story of activity trackers in this chapter is about how some datafication (here, things ranging from step counts to calories burned to sleep cycles) is implemented to discipline wearers. Imposing numerical goals may motivate individuals, but these become coercive when institutions track whether individuals are meeting these goals and using this datafication to decide grades in a physical education course or monitor sleep patterns of student-athletes.

Articulations between wearable technologies and sports have been forged through *gamification*, or the process of turning something that is nongame into a game with rules, points, and rewards. In turning running into a social game—as the first section below explores—gamification becomes established as a potentially pleasurable means to encourage collecting and sharing data within one's social circle, engendering a subject who is well-disciplined not just to exercise but also to datafy exercising. From there, the subsequent sections show how the logic of games has manifested in physical education courses to encourage students to have their performance be tracked and compared against other students. The final major section of this chapter considers how collegiate athletes are asked to embrace gamification and monitoring to optimize their individual performance, and how this acts as an alibi for increasingly pervasive surveillance measures enacted against these students.

GAMIFYING SOCIAL LIFE

To build a better understanding of gamification as a means of articulating wearables and sports, this section begins with the Nike+iPod tracking technology introduced in the mid-2000s before turning to Fitbit, Strava, and a more conceptual development of how gamification operates with technology users and runners. Recording athletic performance for training and feedback has been around for some time,[6] with some sports like cycling having long embraced data tracking as part of the social life of training.[7] Nike+iPod was created via a partnership between Apple, Nike, and a team of biomedical engineers to help track activity—chiefly running—using a shoe chip sensor and an iPod Touch. Available from July 2006, Nike+ positioned the runner as "both the researcher and the subject—a self-contained experimental system. . . . You're actively observing yourself, and just that fact not only provides information you can act on but may also modify your behavior."[8] The Nike+iPod tracker was a small sensor measuring 1.4 by 1.0 inches. Its front face had on it both the Nike "swish" logo and the Apple logo but did not have any kinds of buttons, screens, or interface. Instead, the sensor was placed inside the sole of a dedicated Nike+ shoe, allowing the datafication to measure from the contact a runner's foot made with the ground throughout their run.[9] To manufacture the data, the sensor connected to a dedicated receiver that could be plugged into an iPod Nano. Once the system of sensor, shoe, receiver, and iPod was configured, wearers received "information on time, distance, calories burned and pace" through the iPod's screen display as well as audible feedback.[10] Following their run, wearers could sync the data to a dedicated website to track statistics over time. Nike+iPod was initially promoted to help give amateur runners immediate feedback on their performance. One advertisement from 2007 features a pair of Nike shoes interlaced with headphones, with an iPod placed between the two shoes. The tag line is "Tune your run."[11] The language of "tuning" is a music pun due to its integration with an iPod music player, suggesting a run could be calibrated in the same way as guitar strings. Here, the wearable acts as an appendage for refining physical activity through datafication, using its reports to adjust one's method of running until reaching a state of optimization.

The device used an accelerometer to detect a foot hitting and leaving the ground, a Bluetooth transmitter to send that information to the Apple device, and a battery to ensure information gathering and transmitting would continue throughout the course of a run. By calibrating not only how many steps were taken but also their intensity and frequency, the Nike+ sensor could provide feedback on distance traveled and calories burned and could further track this data over time. While this device could presumably tell runners datafied stories about their own experiences, it also had the capacity to tell Nike stories about these runners. As *Wired* reported in a feature on the Nike+ system: "The most popular day for running is Sunday, and most Nike+ users tend to work in the evening. After the holidays, there's a huge increase in the number of goals that runners set. . . . Nike has discovered that there's a magic number for a Nike+ user: five. . . . Once they hit five runs [with the device], they're massively more likely to keep running and uploading data. At five runs, they've gotten hooked on what their data tells them about themselves."[12] Reproduced here is one of the ongoing promises of datafication: knowing what people do, when, and to what extent to understand behaviors more precisely.[13] If wearables do bring order to life, this form of datafication suggests that order is "out there," waiting to be located and understood through constant datafication of the entire population of Nike+ wearers. As Lisa Gitelman has rightly noted, the idea of "raw data" is an oxymoron; all data are shaped and interpreted through human observation, software, or other forms of measures, statistics, or analytics that give meaning to data.[14] It is up to the engineers, programmers, or spokespersons reviewing these data to unpack *why* they matter and *what* they say.

Wired's feature indicates that sharing running data is not just confined to application leaderboards, friends, and social media; it also entails sharing one's data with corporations, either intentionally or unintentionally. Aggregate data generates "insights" about human patterns of conduct. The quantitative dimensions of habitual running suggest a logic that pays little attention to the qualitative descriptions of this habit; this data manufacturing is not interested in what individual runners do or feel, but rather culls supposed truths from population-level averages of distance and intensity. While individuals may have the power to make sense of their own datafied activity, they are relatively powerless when it comes to how their data are compared to others and used to assess averages and norms. While a generally effusive piece of journalism, this feature in *Wired* nevertheless shows how datafication-based surveillance is framed as understanding something about human behavior patterns. The establishment of norms occurs through data manufacturing. This understanding of behavior is based on the willing choice to participate, thus creating data comprising individuals who desire these devices in the first place.

While Nike+ could track an entire day's movement, it was marketed almost exclusively as a device for runners and other athletes. Fitbit, by contrast, was marketed as a device for anyone to track steps all day, every day, except when the device needed a recharge. Fitbit, from its earliest iterations, was designed to "constantly sense your body's motion. . . . [A]ll this data is collected on the device, it's wireless, so any time you walk by the base station the data is automatically uploaded."[15] When Fitbit appeared at the TechCrunch 50 conference in 2008, journalists were quick to note its potential utility, suggesting that Fitbit "could [do] a lot of us a lot of good."[16] From its beginnings, Fitbit has been entangled with discourses of defining a "good" body and assisting that "good" body with maintaining particular degrees of fit. Despite the development of health-monitoring sensors like those discussed in chapter 1, the Fitbit's major function arguably remains counting steps. The device's popularity, in part, has given rise to expressions like "get your steps in," referring to activities that help wearers reach a targeted step count goal.[17] The introduction has already discussed the arbitrariness of Fitbit's ten thousand steps per day metric, and chapter 1 explained how the Fitbit's coordination between accelerometers and algorithms is more about producing than recording bodily activity. There is an ongoing tension between Fitbit's apparent claims to technical objectivity and the fact that its algorithms generate a representation of humans and their activity based on normative understandings of qualities like gait and pace.[18]

Steps offer one metric that can be made meaningful through their curation and the ensuing computational analysis and social competition they might produce.[19] Datafication can also be shared, such as sharing one's daily fitness summary through the Fitbit application with other "Fitbit Friends."[20] Fitbit Friends act as a form of social capital: the capacity to be admired because of one's physical accomplishments.[21] This is a mode of what Pierre Bourdieu called embodied social capital, in which one's body carries the markers of success and distinction.[22] Wearing the device and sharing its data becomes a means by which to attain and maintain this form of capital.

This embodied social capital operates through the structures of gamification, one of the ways Fitbit incentivizes user engagement in its social leaderboard function. Gamification encourages people to share data with applications and with one another. It is also, importantly, a purportedly pleasurable form of surveillance. For sociologist Jennifer R. Whitson, "play is a cultural practice and public legitimization tool that encourages the acceptance of otherwise contentious technologies. Gamification applies playful frames to non-play spaces, leveraging surveillance and competition to encourage behavior changes."[23] Datafication and gamification work together, relying "on data collection, followed by visualization of this data and cross-referencing, in order to discover correlations, and provide feedback to modify behavior."[24] With the Fitbit, that behavior is physical activity rendered as, at its core, step counts. Gamification invites the wearer to imagine the habituated routine of movement as literally rewarding practices, conferring social status and virtual achievements through climbing the leaderboard. These practices are guided by a "never-ending leveling-up process, guided by a teleology of constant and continual improvement,"[25] in which the game is never really finished. Or, as games scholar Ian Bogost puts it, "Gamification is the goal of gamification."[26]

What gamification highlights, if one takes part in it long enough, is maybe not so much self-actualization, but rather the actualization of philosopher Gilles Deleuze's core tenet for the control society: "One is never finished with anything."[27] As soon as a Fitbit user achieves their goal for the day, they risk becoming bested on the leader board. Each day resets, as do the number boards. Step counts always return to zero. The constant resetting of accomplishments means that social capital is never secure and must be constantly reproclaimed through the ongoing gathering and sharing of data. For quantification to, well, count, the bands implicitly demand to be worn. The very act of "counting"—as in existing or having facticity—intermingles with other meanings of "counting"—as in quantifying or summing up.[28] It is not the measurement of a step that really matters, in other words; it is how the calculation of steps becomes generated toward social visibility and legitimacy.

This visibility can accidentally become connected to broader regimes of surveillance. Consider a much-publicized incident in early 2018, when U.S. military personnel accidentally divulged the location of a secret military base through sharing the running data that the service Strava had generated.[29] Strava is an app for tracking a variety of physical exercises and is compatible with wearables like the Fitbit to provide more detailed metrics and histories of one's physical activity over time. In late 2017 Strava released "Global Heatmap," which its representatives touted as "the largest, richest, and most beautiful dataset of its kind," designed to visualize two years of historical data from Strava users to demonstrate the pathways users took most frequently while syncing activity with the application.[30] Strava was purportedly making it easier for runners to identify popular routes they could build into their routines, as well as generally providing population-level understandings of movement routes. Soon after the release of the map, analysts began sharing findings on social media that demonstrated heatmap pathways ostensibly in the middle of nowhere in countries like Afghanistan and Syria, potentially revealing coordinates of military bases. According to a summary from the *Guardian*: "Zooming in on one of the larger bases clearly reveals its internal layout, as mapped by the tracked jogging routes of numerous soldiers," even though the base itself is not viewable on services like Google Maps. Other locations like a U.S. Air Force base in Nevada were outlined through a pathway tracing the perimeter through at least one individual's daily activity route.[31]

To put it simply, there are consequences to sharing manufactured data, and these consequences are often unanticipated. While Strava offered its heatmap as a "beautiful" visualization of habits and pathways, analysts quickly discovered the map could be used as a surveillance tool that could take advantage of the geolocation of these pathways. Datafying athletic activity sits at a complex intersection between pleasure and power. On the one hand, wearable activity trackers and their attendant applications encourage gamification and social capital.[32] On the other hand, gamification can interpellate the wearer into a system of discipline, in which the device must be constantly worn to manufacture the data necessary to stay in the game. For the theorist Michel Foucault, the emergence of disciplined subjects—individuals who were able to essentially self-govern themselves in accordance with the moral and legal structures of their societies—depended in part on "technologies of the self."[33] In Foucault's theorization, subjectivities are constituted in part through historically situated practices, many of which are socially constructed.[34] Because of the devices' attendant focus on "disciplining" the wearer to be active and hit various goals through their constant use, a number of researchers have suggested wearables are technologies of the self, offering techniques for self-care and self-improvement in line with promoting broader social imperatives to act as a "healthy body."[35]

This mode of disciplinary subjectivity can be empowering in its generation of pleasure for some individuals, much as it can feel disempowering for others. My goal in this section is not to suggest gamification or other modes of sharing means these runners are somehow dupes interpellated into a power structure. Rather, it is to point to how datafying athletic activity depends on disciplining individuals toward becoming data-producing subjects. As the example of the Strava heatmap demonstrates, these subjects may accidentally divulge data such as location. The next two sections explore discourses around physical education and student-athletes to further consider how datafication operates as a form of surveillance.

STEPS FOR GRADES

In January 2016 Oral Roberts University—a private university in Oklahoma—announced it was implementing a mandatory use of fitness trackers for incoming students. According to news reports following the university's announcement of this program, ORU had required a physical fitness course for all undergraduates for many years. Students enrolled in this course had traditionally been asked to perform 150 minutes of activity each week and keep a log of their activities in a journal. This activity log would then be submitted to an instructor at the end of the semester as part of their grade in the course. The incorporation of Fitbits significantly altered the dimensions of the course, as students were asked to purchase Fitbits (which retail at roughly US$100) as part of the costs of the course. The trackers would be used to monitor an array of activity metrics, such as steps taken, distance traveled, calories burned, and heart rate.[36] The university insisted it would not monitor students' weight or diet (data that can be manually provided on Fitbit's website but is not automatically generated through wearing the devices). One year after the program was implemented, ORU reported it had collected "more than three billion steps," and that nearly thirteen hundred students were wearing the devices in the spring 2017 semester.[37]

Whereas the previous section suggested individual self-tracking might indicate how Fitbits act as "technologies of the self" and further entrench modes of disciplined subjectivity, ORU's report indicates how the administrative use of wearable technologies to monitor activity levels of a population is in line with what Gilles Deleuze theorized as "societies of control."[38] Deleuze, building from Foucault's work on discipline, suggested in the early 1990s that an emergent logic of "control" was taking shape, whereby technologies and economies would become increasingly fused together to, among other things, monitor the comings and goings of people across the spaces and transactions of their daily lives. In the previous section I alluded to Deleuze's proclamation that in control societies, "one is never finished with anything" due to constant imperatives to either input data such as passcodes or provide transactional data in economies. Also important in Deleuze's sketch of societies of control was the concept of the *dividual*, mentioned in this book's introduction. To recap: where previous modes of administration might have considered "the individual" to be the lowest unit of a population, Deleuze foresaw how control societies would work to further "divide" an individual into increasingly discrete units of measurement.[39]

When physical activity trackers are used to evaluate student behavior, they are operationalizing the dividual, treating each student as a series of metrics to calculate throughout a semester to render a grade (itself a metric of performance with consequences for scholarship eligibility, degree requirements, and the like).

This program is an example of a mandate, in which individuals are compelled to use a wearable to serve an institutional goal. Perhaps accidentally hinting at the actual purposes of the Fitbit program, a schedule for ORU's fall 2018 new student orientation informed incoming students that as part of their registration they would "receive [their] ORU I.D. and Fitbit."[40] The Fitbit came to be a crucial component of a student's identity—whether they realized it or not—because ceding data to the university provided it with a sample size of thirteen hundred students to run populational analysis, determine averages, and see where outliers were located. Under the guise of promoting well-being, universities are developing the capacity to learn about the everyday mobilities of undergraduate students. For Foucault, power structures produce effects; the effect here is to constitute students as data-generating subjects participating in a "game" of the university's design.[41]

The Fitbits outsource the work of logging and ensure students are, in fact, completing requirements. As news reports have detailed, once students authorized the university to access their data, Fitbit could send reports automatically on behalf of a student to the university's central server. Jeff Stone, in the *International Business Times*, suggested this was not something to worry about, as "Fitbit activity makes up 20 percent of a student's grade for a single education class, which all students are required to take during each semester at the school. The Fitbit replaces a previous program where students entered their fitness activity by hand, and students can opt out of the program *in favor of a slightly lower grade*."[42] While Stone made the case that the ability to opt out of Fitbit data collection rendered the criticism surrounding ORU's implementation of the device a nonissue, the grade punishment signifies that there was a cost to opting out of datafication. Students did not truly have a choice—especially those, like many undergraduate students, who were focused on maintaining high grade point averages (itself a numbers game)—when one choice involved a penalty and the other did not. This is one way institutionalizing forces try to game the implementation of wearable technologies: they make them "not required" while also demonstrating how using a wearable is the only way to achieve the highest outcome. Moreover, students are—in this model—not encouraged to learn how to log and interpret their own data, but to allow the institutional representatives to interpret it on their behalf.

In ORU's program, Fitbits are not about self-improvement; they are instead about datafying and analyzing population behavior.[43] ORU's decision to do away with student journals and replace them with Fitbits suggests that humans do not produce reliable data without the intervention of an intermediary. Fitbits become figured as supposedly impartial agents capable of monitoring undergraduate students through coerced datafication.[44] While ORU's Fitbit program received a good deal of national attention when it was announced, such projects are increasingly prominent in high schools and middle schools. These programs use a variety of specially designed fitness trackers that allow physical education teachers, parents, and other administrators to view and assess student activity both during class and remotely. Hidden Valley High School in Roanoke, Virginia, began implementing Polar A360 wearable trackers during its physical education courses in 2016, the same year ORU began its Fitbit program.[45] These trackers have many of the same functions as the Fitbit, but they operate in a system that allows the physical education instructor to set goals for students and gather real-time data about aerobic performance. The Polar devices look like conventional fitness trackers; they are worn against the wrist and have a display screen showing the time and a readout of the data being generated through the device's sensors. The Polar wristband monitors heart rate and puts students in the "green zone" if their heart rate is between 70 and 80 percent of their maximum heart rate, based on height, weight, and age inputs that the students provide. According to news reports on this program, students must meet these goals to earn participation points.[46]

The Polar devices are connected through the GoFit system, a software application that allows physical education instructors or other administrators to view activity data in real time.[47] In addition to using heart rate data to grade students on whether they have adequately completed their day's physical education requirements, the Polar GoFit system also puts students in competition with one another. According to reporter Sara Gregory, the Hidden Valley auxiliary gym contains a screen display "showing students' names and the number of minutes they'd spend in moderate, hard, and maximum heart rate zones. A color indicate[s] their current zone."[48] In asking students to compare their heart rates to one another, the metrics can be both aspirational—in encouraging forms of social competition—as well as coercive, revealing to the class which students struggle to reach prescribed fitness levels. Polar's website frames some of the technical stakes and processes for using the GoFit system. As its privacy policy details, GoFit processes all the data the school provides. Each school "is responsible for what information is saved in the system and how it is handled. The school is also responsible for the accuracy of the students' information and requests from individual students (such as requests to delete data)."[49] The GoFit system also allows teachers to track assessments and share data with parents (who must provide consent for their children to use the device) and administrators. As at ORU, the articulation between wearables and physical education is enforced rather than voluntary.

The chapter moves in its next section from more general discussions of datafying physical activity to some particular articulations between wearables and student athletes as a more pervasive form of surveillance.

OPTIMIZING AND MONITORING COLLEGIATE FOOTBALL PLAYERS

Wearables have increasingly been implemented in university athletics, where companies like Catapult and Whoop have become a veritable cottage industry, marketing themselves as means for coaches to manufacture understandings of players during each practice. The Catapult One is a small tracker that is placed inside a dedicated vest. The tracker itself is not wearable but relies on being tucked into a slot on a form-fitting vest, resting on the back between a wearer's shoulder blades. The tracker additionally has no buttons at all; it is a dark grey oval about the size of a human palm. When the tracker slides into the vest, it detects that it has been inserted and begins producing data. The data created through the tracker is sent to a dedicated application through which athletes and coaches can review practice and training sessions in attempts to learn more about athletic behaviors. Metrics that Catapult One produces include total distance, top speeds, and sprint distances. Catapult One tracks movement in ways somewhat like more popular devices such as the Fitbit, but it is framed for athletes and coaches who are focused on different intensities of movement. Whoop, by contrast, looks more like a Fitbit or an Apple Watch: it is a wearable band that rests on a player's wrist. It does not have a graphical interface of any sort; like Catapult One, it is a tracker designed to rest against the body and measure levels of activity and bodily function like heart rate. The Whoop application retrieves the data produced on the wearable and provides reports on things like sleep, recovery, heart rate, and respiratory rate, but accessing its reports as well as its automated analysis and recommendations on how to interpret these reports into actionable suggestions costs, with the Whoop 4, US$30 per month.[50] Catapult and Whoop demonstrate how wearables for athlete training do not rely on the touchscreen interfaces of smartwatches; instead, they prioritize stuffing sensors into the wearable trackers and offloading data reports onto dedicated smartphone applications.

Other wearable trackers are explicitly focused on measuring sleep, such as the Rise Science sleep system.[51] Rise Science is an application that uses a smartphone or wearable's existing sensors to monitor and measure sleep. In its partnerships with athletes, the company has also offered dedicated bed sensors that can track bodily movement.[52] The application provides graphical layouts of various dimensions of sleep, such as charting planned and actual levels of sleep, which are accessible to both players and coaches. Such systems have come with their own concerns about the monetization of player data, especially for student athletes who are not compensated for their performance.[53] However, these wearables are more often framed as devices for helping players. As Clemson football coach Dabo Swinney said in 2017 to justify tracking the sleep of the school's players: "It's kind of like back in the day, when we were having nutritionists, nobody wanted to talk to the nutritionist . . . but now it's cool to be with the nutritionist. . . . So we're trying to create that same culture with our sleep, and these guys understanding how important it is for their performance."[54] Swinney's comparison falls apart rather quickly upon examination: "back in the day," nutritionists were not automatically gathering eating data from players about when they were eating, for how long, or which foods. Yes, players might eat prepared meals at dedicated team dinners, but nutritionists could not monitor what a player chose to eat in their own dorm or apartment. With Rise Science, players either wear or strap a device to their beds to manufacture data on their sleeping. This is a culture of surveillance masquerading as a culture of wellness.[55]

This section focuses on college football in the United States as a high-profile and highly profitable industry. As such, its players are valuable assets whose success in games contributes not only to the reputation of the university, but also to ticket sales, alumni donations, and other means of generating revenue through licensing and media deals.[56] Some universities, like Auburn University in Alabama, have hired private security firms to monitor student athletes.[57] At other universities, "class checkers" are paid to make sure student-athletes attend the classes in which they are enrolled.[58] This is to say nothing of the various social media policies at universities, some of which involve monitoring player accounts and some of which prohibit players from using public-facing social media during a season.[59] Student-athletes in general and college football players in particular have been a highly surveilled group since before wearable technology arrived.[60]

As much as this technology extends the capacity to monitor and datafy sleeping activity, it also helps create additional sources of value and capital. For example, in 2016 the University of Michigan began a collaboration with Nike that drew criticism: the contract "could, in the future, allow Nike to harvest personal data from Michigan athletes through the use of wearable technology like heart-rate monitors, GPS trackers and other devices that log myriad biological activities" and gave Nike the broad authority "to utilize" the data it had collected.[61] While Michigan and Nike both, naturally, assured worried members of players' associations and regulators that appropriate anonymization would occur, the example underscores that datafication is a valuable industry, in terms of both the knowledge it supposedly provides for players and the additional research capacities, product development, and, yes, revenue it could provide whoever retains the rights to the datasets.

Despite these ongoing concerns, devices like Catapult and Whoop have made significant inroads into the biggest NCAA football organizations and have been regularly touted in local and national sports journalism as key tools in athletic performance. The Catapult GPS and performance-tracking wearable was credited for the success of multiple national-championship-winning NCAA football programs in the 2010s. After Florida State University won the championship in 2014, ESPN pointed to Catapult as a means of generating motivation among team members during training. Student-athletes perceived it as a game: "the benchmark for a juggernaut for which the biggest challenge comes by competing against itself," or in which players constantly check against each other's data to try to outperform each other.[62] For coaching staff, Catapult's datafication offers strategies for optimizing practices and analyzing performance metrics. As Florida State head coach Jimbo Fisher put it in 2014: "It's not the reason you win. . . . But it takes a lot of the guesswork out of how your team is feeling, how individuals are performing and how you moderate practice."[63] There are again shifting modes of knowledge production at play: coaches feel less sure of established heuristics and position datafication as an objective rendering of an athlete's performance, allowing them to make decisions.

After the University of Alabama won the 2015 national championship, some of its athletic staff also pointed to Catapult as an important part of helping players make explosive plays and giving coaches datafied renderings of player performance. According to Jeff Allen, the associate athletic director of sports medicine at University of Alabama, "it just provides us with tremendous amount of data that helps us to make performance decisions. It even helps us to make some medical decisions," such as understanding when players may have overexerted.[64] According to a feature on AL.com celebrating the transformative impacts of Catapult, Alabama's football coaches and training staff focus on data from the device's odometer (how far a player runs in a given day), its calculation of maximum velocity (how fast a player runs), its measurement of "explosiveness" (how quickly players change any given direction), the speed yards of a player (yards run over twelve miles per hour), and player load (or exertion over time).[65] Catapult straddles the spaces of rehabilitation and optimization, where trainers can use it to monitor players in recovery while also promoting extraordinary data points.

By 2021 Catapult was priding itself on the championship-level results its product was ostensibly generating: "All four teams in the College Football Players are clients, and all five champions from the ACC, Big Ten, Big 12, Pac-12, and SEC collegiate conferences are using Catapult solutions." While coaches and athletic staff emphasized student-athlete safety and injury reduction in their public comments on Catapult, CEO Will Lopes directly suggested Catapult helps win games: "There is clearly a competitive advantage afforded to teams using our solutions, especially those using video analysis combined with our performance data."[66] This is, again, an extended version of sabermetrics, one that suggests datafication is key to organizational optimization and the production of winning seasons.

Whoop is a similar wearable technology designed to assist athletes with training and health. Penn State University announced its partnership with Whoop at the end of 2022, saying that "every student-athlete will have the option to wear WHOOP 4.0, participate in onboarding and ongoing in-depth education on how to optimize their performance and gain a deeper understanding of key trends across sleep, recovery, and strain."[67] Other teams such as the University of Miami have suggested data manufactured through Whoop helped identify when early-morning workouts were actually hampering student-athletes because they were chronically undersleeping.[68] It is telling that it took datafication for these coaches and staff to make these decisions, especially since overexertion and "playing through the pain" have been recurring problems not only for collegiate football but for nearly every form of organized sports.[69]

In promising to solve these problems, however, these reports fail to acknowledge the additional privacy risks at play in using these technologies. In a sports law review of data collection through wearable technologies, Alicia Jessop and Thomas A. Baker identified utilization of invalid or improperly interpreted data as well as privacy concerns about the use of these data over time.[70] For instance, there have been limited validity studies of these technologies beyond company white papers or the anecdotes described in some reporting. Both Florida State and Alabama noted learning curves in interpreting data and results that did not seem to align with observations from coaches. Datafication, in other words, does not grant access to an objective reality. As with nearly all the examples discussed so far in this book, it actively constructs a reality for those with some form of administrative power to interpret.

Health monitoring—be it during practice or charting athletes' sleep—becomes an alibi for surveillance and control. While players may perceive it as a pleasurable gamification strategy, this process also encourages players to see surveillance as normal, to quite literally sleep while manufacturing data for coaches to review. As head coach for Pennsylvania State University James Franklin put it:

> Basically, they sign off on it, and we can see: are they getting enough sleep? What type of sleep are they getting? . . . Are they getting the most value? It's not something we require them to do. They choose to do it. . . . Now instead of just looking at their Twitter or whatever it is, you can actually see, y'know, he went to bed at 11 o'clock or he went to bed at 1 o'clock or he went to bed at 11 and he tossed and turned all night long. Why did he toss and turn? Does he have sleep apnea, what is it? Now we can study that and help him because I think we all realize the better sleep you get the more your body is going to recover and the more productive you're going to be the next day.[71]

It is worth quoting Franklin at length here, as this interview condenses much of what this chapter has emphasized. He acknowledged that, on the one hand, Whoop replaces existing surveillance mechanisms like "Twitter or whatever," with which coaches can monitor players' online activity. Whereas social media requires public disclosure and a choice to participate in sharing photos, liking accounts, or other forms of engagement, Whoop manufactures the data automatically. While Franklin acknowledged its use is voluntary and the players "sign off on it," it remains unclear what amount of informed consent is provided to players, especially given the concerns around player privacy, ongoing data collection, and validity mentioned previously. Finally, he pivoted from surveillance to focusing on health benefits, such as a hypothetical case of sleep apnea, which only becomes apparent because of a player having restless sleep. In these ways, the device is framed as a kind of preventative health solution, and surveillance is framed as a mode of care.

Surveillance and care have often been linked in health communication and health care when discussing vulnerable populations.[72] For college football players in particular, overexertion and conditions such as rhabdomyolysis join risks such as concussions, ACL tears, and a variety of other short- and long-term injuries marking these student-athletes as, at least to a point, a vulnerable group, in that the labor that puts their bodies in a high-risk situation is tied to revenue dollars for the school, a scholarship for their education, and any potential professional aspirations and possible earning income. If devices like Whoop and Rise Science help care for players, then it is only because they are proposing to solve a problem that coaches have created and maintained within the current operating conditions of college football itself. Rather than systemically reforming the game, the datafied surveillance proposes to maintain the status quo until a problem emerges.

It is perhaps better to suggest wearable technologies in collegiate sports are about maintaining a political economy of student athletic performance. Putting the onus on the device to alert coaches to player conditions absolves the athletic trainers, at least in part, from modifying their workout schedules until an issue arises. Players are tracked beyond the practice field in the name of preserving their care and in the name of coaches being concerned with their well-being and their "productivity," but this masks how—at least for the high-profile programs mentioned throughout this section—student-athletes are also assets. That Franklin equated the need for sleep with "value" points to a need for student-athletes to retain (and improve, if not optimize) their value while affiliated with the university. Student-athletes may see the use of the devices as a fun and interesting way to learn more about how their bodies perform in different conditions, but the gamification aspect—seeing it as a challenge to keep one's condition strong to improve metrics over time—again reveals datafication's administrative project of giving more control to those with authority over student-athletes.[73]

CONCLUSION: GAMIFICATION IS CONTROL

Across amateur runners, physical education students, and student athletes, discourses around the articulation between wearable technologies and sports have increasingly promised increased optimization via increased monitoring. Studying the articulations between wearable technologies and sports offers one way to glimpse the slippages between datafication as a means of ordering everyday life for personal improvement and motivation and datafication as a means of ordering everyday life for surveillance and control. The gamification that characterized Nike+iPod and the social sharing of data among runners helped to set the conditions for datafication to be perceived as a valuable, helpful, and frankly fun tool for athletes to use. These perceptions helped make wearable technologies appear useful for physical education teachers and collegiate athletics departments alike, who could use the datafication and aggregation tools to motivate students, monitor populations, and—for college student-athletes—detect possible issues with workout load and thus prevent injury. While this combination of gamification and datafication is undoubtedly enjoyable for some young people and athletes, a focus on the pleasurable side of gamification distracts from surveillance creep.[74]

While this chapter has focused on amateurs and students, I end it as I began it, with Major League Baseball. The MLB approved the use of some wearable technologies during games in 2016. If *Moneyball* demonstrated an epistemic shift in professional baseball, the development of technologies that can assist with datafication and analysis during training and gameplay has continued to expand. Gleaming profiles in publication outlets like *The Athletic* have touted wearable harnesses to improve pitching and hitting: "I'll show you one of the best players in the game right now . . . and I'll show you wiggle room," promised Steve Johnson of the Advanced Baseball Training and Performance Center.[75] This celebration of the athlete as a kind of data well, gushing data into analytics software, harbors concerns about how this data can be used and by whom. In a review of issues stemming from the agreement to allow wearable technology into baseball, John A. Balletta suggested the data could influence player value during trading or interfere with contract extensions, or that the data could be used to release a player if they can be deemed a health risk for a team.[76]

While in some ways this chapter carries through lines from chapter 1 in its focus on the logics of health, wellness, and injury reduction used in the promotion of these devices, it also demonstrates again how forms of public discourse work to legitimize these devices and accept their implementation as necessary and good, focusing on optimization and care over surveillance.[77] In examining tensions between individual motivation and institutional control, this chapter has considered how the institutional deployment of wearables in physical education courses and collegiate sports organizations enacts disciplinary surveillance and control under the guise of care, health, and, yes, winning games. The next chapter of this book pivots to engage with wearables in workplace surveillance and how the tensions between pleasure and power slip away quite quickly in nongaming or nonathletic contexts.

FOUR

Labor

OR: WORKPLACE SURVEILLANCE DOWN TO THE MILLISECOND

"IF WORKERS SLACK OFF, the wristband will know." So warned a headline in the *New York Times* in February 2018 about a patent Amazon had recently filed,[1] describing a bracelet that could mediate between a warehouse worker's body and a variety of inventory bins. Through coordinating with ultrasonic sensors, the bracelet could help a worker locate items efficiently while a related "management module monitors performance of an assigned task."[2] In its coverage of the patent, the *Times* interviewed Max Crawford, a former Amazon warehouse worker who described his experience of working for the company in the following way: "They want to turn people into machines."[3]

Throughout that first week of February 2018, nearly every major news outlet in the United States reported on Amazon's wearable patents. Heather Kelley at *CNN* speculated, "In theory, a company could fire a worker if their wearable found they were performing tasks slower than co-workers."[4] Amazon's representatives challenged similar speculations at popular technology website the *Verge*, claiming that "by moving equipment to associates' wrists, we could free up their hands from scanners and their eyes from computer screens," while also adding for good measure: "Like most companies, we have performance expectations for every Amazon employee and we measure actual performance against those expectations, and they are not designed to track employees."[5] The response here is telling: Amazon has performance expectations for each employee, even though the devices are not designed to "track" employees, while at the same time it meant to make employees more efficient. It also seems to confirm, at least to a point, Crawford's hunch: that Amazon's pursuit of maximizing human labor would deploy wearable appendages and an array of sensors to make human workers more like the robots that maneuver across warehouse floors to help complete orders.[6]

The Amazon patents gained particular attention because of reports about poor working conditions in Amazon warehouses related to long hours, unreasonable turnaround times for fulfilling orders, a need for stronger benefits and more secure jobs, and a lack of breaks and rest, among other things.[7] Of course, while the publication of a patent does not guarantee Amazon was actually manufacturing such wearables (as journalists and representatives were quick to point out), their existence nevertheless participates in Amazon's ongoing imagination of its workforce as machinic.[8] Amazon's warehouses have served as a focal point for debates about the robotization of the human workforce and the use of digital tools to broadly surveil and evaluate employees; the company is but one strand of an increasingly complicated web of employee surveillance across many different workplaces. Surveillance of email, keystrokes, and internet browsing trains employees to accept their actions are routinely monitored, while other workplaces have replaced door keys with identification cards capable of tracking how employees enter spaces and use products like printers and vending machines.[9]

This chapter explores worker surveillance as a means of understanding the articulation between wearables and labor, focusing on the technology company Humanyze's sociometric badge. It analyzes how the badge has been imagined and represented in news reports, academic studies, and promotional talks by its president, Ben Waber. Through this analysis, this chapter attends to how wearables are pitched as useful guides for providing feedback on a variety of efficiency- or productivity-based metrics.[10] Waber invokes a desire to order work down to the millisecond, analyzing all worker actions through datafication to, supposedly, optimize work performance. This chapter shows how sabermetrics extends beyond the domain of sports and has been explicitly used to surveil office labor. The discussion of Humanyze furthers this book's engagement with surveillance as a mode of discipline and control that is often framed as a mode of caring for and improving working conditions.

SOCIOMETRIC ANALYSIS AND AGGREGATE METADATA

Ben Waber began developing what would eventually become Humanyze as part of his graduate studies at the MIT Media Lab. As the introduction to this book mentioned, the Media Lab was an early supporter of wearable computing and hosted some of the earliest symposia to encourage development and public recognition of a variety of potential wearable technologies. As a site for "Inventing the Future," the lab specialized in a determinist ideology that trusted computer technology held the key to solving many of the social and cultural issues of the twenty-first century.[11] For founding director Nicholas Negroponte, this was perhaps best described in the guiding ethos of his 1995 book *Being Digital*, which confidently proclaimed, in describing the unstoppable march of computer-based cultures: "The change from atoms to bits is irrevocable and unstoppable."[12] Technology historian Morgan G. Ames has examined some of the consequences of Negroponte's unflappable faith in computer technology in a study of the lab's One Child One Laptop program, which proposed to change education through giving hundreds of laptops to schoolchildren around the world, without providing adequate infrastructure, education, or training.[13] The result, as Ames details, was a policy that failed to consider how to make technology work beyond simply providing the devices. Issues of literacy, competency, and technical training were not part of the One Child One Laptop program, making its actual impact highly variable, and in some places it was completely ineffective.

Humanyze is part of the Media Lab's commitment to technology in and of itself as the arbiter of change. Where One Child One Laptop focused on education efforts, Humanyze focuses on workplace practices. As Waber explained about his graduate research: "When we were at MIT, we started to realize that people were actually already carrying devices that were measuring an awful lot about what they do. . . . If I put RFID [radio frequency identification] readers in the ceiling, I can tell where people are."[14] On one level, Humanyze can be understood as an infrastructural project. As communication researcher Jordan Frith has defined it, RFID "is a communicative mobile technology that can uniquely identify and sort billions of objects and turn various physical processes into trackable data."[15] Crucially, these processes of identification and sorting occur largely through machine-to-machine communication embedded in transactions in badges and sensors.[16] RFID is part of what makes worker surveillance possible with Humanyze, and it is also a major component in other wearables like the Walt Disney World Magic Band, discussed in chapter 6 of this book. For this chapter, it is enough to suggest that Waber and his team originally saw the utility of RFID as a means of tracking humans without interference.

Claims to measure people through the devices they carry suggest that—like the heart rate monitors explored in chapter 1 and the sleep trackers discussed in chapter 3—that machines have become superior arbiters of what humans do and know. In Waber's writings and his interviews about the device, he described Humanyze's origins in MIT's Human Dynamics group to create what they called a "sociometer," which would combine multiple sensors into a single wearable box: "The idea was to make a general ID badge that would be able to measure all the different signals—[infrared], motion, and sound—at the same time."[17] These sensors included radio frequency identification, which could interact with other machinic sensors in workspaces; infrared transceivers, which could detect when two people wearing the badges came into proximity to one another; accelerometers to track movement over time; and microphones to analyze vocal tone and conversation length.

The sociometric badge, according to Waber, is "the natural evolution of the company ID badge. No longer just a tool to open doors, this new kind of ID badge enables you to understand yourself and your company at large through data-driven reports and feedback."[18] The badge is connected to a lanyard draped over the wearer's neck. It is meant to feel like wearing an employee identification card around one's neck. It has no actual interface for the wearer to engage with, but the interior of the badge houses a range of sensors, as listed previously. There is a charging port on the bottom, and there are three small black slits on the front—presumably for the microphones and sensors to better manufacture data—but there are no buttons and no layouts, just an off-white slab. In other interviews, Waber emphasized the inability of the microphones to record audio: "The microphones in the badges only record about 48 samples per second, which can track tone of voice and volume but not content."[19] Per hour, then, each Humanyze badge can use its microphone to record close to 173,000 audio samples that, while they may not be analyzing content, still record a substantial amount of chatter to be analyzed. As Waber described it in a 2019 interview with *MIT Technology Review*, the incorporation of microphones and audiological analysis stemmed from early research in the Media Lab:

> And the question there was, based on how people spoke, not words but just the way that people talk, could you predict, for example, who's going to win a negotiation. And you could, really accurately, with about 85 percent accuracy . . . [t]one of voice, interruptions. The body language as well really does matter. . . . Typically what you do to study those kinds of features is,

> you'd have some poor grad student record a video of an interaction and then they'd have to go frame by frame [through] the video, over weeks for a single conversation, to figure out what was going on. But what we could do now is with sensors and with obviously more and more powerful algorithms, we could actually process that in a matter of seconds.[20]

The lack of an interface and the transformation of spoken voice into audio fragments encapsulate the idea that Humanyze is meant to datafy what workers do throughout their day without giving the worker an opportunity to respond to or engage with the data being created based on their activities.

Waber's promises in 2013 align with other imaginaries and discourses traced across this book, in which claims to know your heart and tune your hearing are supposed to provide individuals with the means to understand how their bodies operate in relation to the environment around them. Similarly, Humanyze supposes that humans, by and large, are not very good at being advocates for their own experiences. Rather than trusting managers to talk to employees and "experiment" with different managerial practices, the company operates as an intermediary to experiment on workers, improve company productivity, and increase profits. The company's name is telling in this regard. The vowel-shifted play on the word *humanize* suggests making something friendlier to humans. To humanize something is not only to make it more humane, but to administer a degree of civil order in otherwise unruly behavior. *Humanize* is a verb; it describes a process of making conditions more civilized. But what, with Humanyze, is being humanized? Are working conditions made more humane (that is to say, courteous, friendly, tender, or better befitting human beings)? Is the company about making worker surveillance appear more palatable (as in surveillance as the process of building civilization in a workplace)? The mission statement, which vows to help companies "uncover how work *actually* gets done . . . decide on the types of interventions to make, and *measure the impact* of change management initiatives on productivity and performance," is not particularly sympathetic to human workers and positions itself more as a tool for management to intervene in the governing of workers than to humanize the conditions of that work.[21]

The first page of Waber's 2013 book *People Analytics*, in which he established his larger ideological goals for the incorporation of data analysis in organizational practice, referenced the Oakland Athletics baseball franchise. In his public presentations about Humanyze, Waber has often begun with an anecdote about the Oakland Athletics, and unpacking this reference is key for assessing the larger epistemological claims at the heart of Humanyze's project. While chapter 3 summarized sabermetrics in greater detail, Waber has framed the management of a baseball team as effectively like the management of office space, and his conceptualization of "people analytics" is perhaps best described as an effort to port sports statistics into the workplace. After beginning his public lectures and presentations with the example of the Athletics, and how they used "data about behavior to build [their] organization," Waber then often asked, "Could we do the same sort of thing for business?"[22]

To answer that question, Humanyze operationalizes several different kinds of datafication to perform different analytics: "Each badge generates about 4GB of data per day, which gets uploaded to the cloud where Humanyze breaks down the 40 types of information it collects and pares it down to the 6 most important pieces for the client. They then link that to a key business metric dashboard, so the customer can see how behaviors are affecting the bottom line performance of the company."[23] Humanyze is a fundamentally asymmetrical form of datafication: while workers generate gigabytes' worth of data throughout their working days, they are not granted the opportunity to review, clarify, or discuss any of that data; rather, it is up to the Humanyze team to analyze and synthesize that data for managers and supervisors to review. Further, the suggestion that these behaviors "are affecting the bottom-line performance" suggests a correlation between this datafication and profit. Waber has insisted that Humanyze's device and software do not *record* the audio gathered through the microphones. Rather, "in real time, there is voice processing, so we're looking at, again, how much I talk, where I spend my time. And that data is always aggregated."[24] While Waber and his team have positioned people analytics as an epistemological benefit of using wearable sensors in the workplace (producing ostensibly "new" forms of knowledge about working practices), public reporting on the product has routinely highlighted fears of worker surveillance and privacy concerns.

The Humanyze team provides a detailed list of protocols they follow to attempt to make sure individual data remain private. Chief among these, repeated throughout interviews, is that managers are only given access to the aggregate data of all employees. The claim is that aggregation is not an invasion of privacy because an individual is only understood as part of a population.[25] But this only works to a point. One of the hypothetical case studies Waber has often mentioned in his interviews is to ask how much the management team talks to the engineering team in a company. As he has described it, this is a problem that most people don't know the answer to: they can't quantify and measure how much members of management talk to members of engineering. In a company with hundreds or thousands of employees, the relationships between different departments might be reasonably analyzed in ways that could protect managers from using Humanyze's data to deduce things about individual workers, but the same scrutiny does not necessarily hold in companies with, say, fifty employees that include a marketing department of four people. As journalist Chris Weller reported: "Waber reassures me the data they get in contains no identifying information. It's all meta-data. For extra privacy, companies themselves can't even see each employee's individual results, only the aggregate webs. And to avoid any claims it's just another Big Brother, Humanyze will only outfit employees who have given their consent."[26] Waber made these claims—which journalists have reproduced without critically questioning—despite studies that have demonstrated how supposedly anonymized and aggregated datasets can be "reidentified" relatively easily.[27] These include studies coauthored by Alex "Sandy" Pentland, the MIT professor who helped develop Humanyze in his Human Dynamics Group (and who made the keynote speech at the 1997 symposium that began this book's introduction).[28] The suggestion that "it's all meta-data" is meant to ease worried employees, who are nonetheless asked to consent to their own performance becoming part of a dataset, their own actions being used as the basis of an analysis of company performance recommendations.[29]

THE OPPRESSIVE SCIENCE OF HUMAN WORK

These conceptions of a wholly observable and measurable workplace are not new. At the turn of the twentieth century, Henry Ford, Charles Taylor, and other figures in the industrial modernization of American business developed the field of "management sciences" to surveil workers and maximize their capacities to work. Perhaps most famous is the refinement of assembly line labor in the first part of the twentieth century, which Charlie Chaplin famously satirized in his 1936 film *Modern Times*.[30] Humanyze extends management science to datafication of labor through wearable technologies. This datafication offers a set of strategies for knowing about the lives of workers through the kinds of data collected about them and analyzed in the aggregate. Management science has always turned its eyes toward technologies that can mediate, calculate, and render some representation of what a human does.

Where Waber has celebrated the use of sabermetrics in baseball as a model for the workplace, historian Nikal Saval has seen the logic of *Moneyball* stemming from experiments in social programming from a century earlier. Saval drew a direct line from statistically measuring the performance of baseball players to the development of efficiency monitoring in the early twentieth century under the rubrics of Taylorism.[31] Taylorism was the shorthand name given to scientific management, in recognition of Frederick Taylor's work in managing human workers as a form of science. As organizational researcher Phoebe V. Moore summarized, Taylor and his followers believed "exact measurement would both eliminate any unnecessary waste of spent time and human effort as well as any possible loss of expenditure for a company."[32] Of course, as Moore explained, Taylorism's efficacy in bringing about meaningful change was contested and unequal, but the assumption that human behavior can be understood through data collection and analysis has remained a powerful ideology for ordering workplaces and monitoring laborers.[33]

Melissa Gregg has charted similar histories through her analysis of the Gilbreths and their time and motion studies, which further sought to analyze different degrees of worker performance, such as how worker bodies were moving in spaces, using stopwatches and timetables.[34] Humanyze extends this, asking workers to understand every part of their day as calculated, from how often they are at their computers, to how they respond to email, to who they come into contact with and for how long. The datafication of labor is framed as a protection of their privacy even as it intrudes into nearly every operation of work. In focusing on aggregation, Humanyze tries to prove the point of the early management scientists: the individual worker does not really matter, insofar as they are understood as a unit of datafication, part of an aggregate to be analyzed rather than an individual to understand.

Efficiency experiments with the stopwatch, designed to regulate employee activity down to the second, are only one facet of a larger history of the intersections between technology, time, and control.[35] Historian E. P. Thompson, for instance, has charted a shift from "task-orientation" to "time-orientation," wherein workers were placed less in relation to the tasks they completed (how many tables were produced) and more in relation to how expediently they could complete work (how quickly a single table can be produced).[36] Thompson mapped the emergence of "time-discipline" from as early as the 1700s: "In all these ways—by the division of labor; the supervision of labor; fines; bells and clocks; money incentives . . . new labor habits were formed, and a new time-discipline was imposed."[37] This form of disciplinary power still lingers through the processes of workplace surveillance, in which the tracking of time as a variable of data analysis is figured as an important metric.

Political aestheticist Esther Leslie has historicized the development of human resources as a "quasi-science of personnel management" related to the management sciences of Taylorism during the early twentieth century.[38] Digital services like Humanyze use "granular analysis" through data mining to facilitate "a science of atmosphere, of corporate mood, of office ambience."[39] The scientific study of efficiency was always about surveilling workers' bodies: how they moved as well as the belief that those movements could be changed and relearned. This is not a science of the worker; it is designed for the manager to learn to monitor the body in a way that economically benefits the company.[40] Taylorism justified the creation of a managerial class for assessing workers.[41] My goal in this section has not been to provide an exhaustive overview of Taylorism, a task the historians mentioned here have done quite capably, but rather to review it briefly in order to situate Humanyze's people analytics project—and the articulations of wearable technologies to labor—as a transformation of management science.[42] In a review of workplace surveillance, labor historian Kirstie Ball argued, "Any discussion of workplace surveillance begins with the idea that surveillance and business organizations go hand in hand, and that employee monitoring is nothing new."[43] By treating surveillance and organizations as "hand in hand," Ball implicitly played out one of Michel Foucault's key tenets: when ideas come to be seen as without history, they become taken for granted and assumed parts of daily practice.[44]

Consider a study that appeared prominently in *People Analytics* and has often been cited in public discourse on Humanyze: the Bank of America call center study. For this study, as Waber narrated, Bank of America contracted Waber and his MIT colleagues to study call center performance: "To study these teams in detail, we collected not only badge data, but also performance metrics, demographic information, survey data, and e-mail records."[45] Waber claimed that thousands of hours of badge data and thousands of emails allowed his team to reconstruct employee interactions down to the "millisecond." In the version of sabermetrics *Moneyball* touts, and which Waber treated as his guiding ethos, the measurement of discrete categories of performance is limited to things that take place during a game of baseball, such as on-base percentage. However, as demonstrated in the previous chapter, organizations like college athletics are increasingly interested in what athletes do while they sleep, extending datafication beyond the field or the practice facility. While the workers in the Humanyze studies relinquish their badges at the end of each day, the workplace itself becomes coded as a space of totalized surveillance. It goes beyond assessing time taken to complete individual tasks, instead considering an entire amalgamation of habits and routines taking place "on the clock" that comprise the working population's daily practice. In Ifeoma Ajunwa, Kate Crawford, and Jason Schultz's terms, this approximates "limitless" worker surveillance.[46]

A *Fast Company* article about the magazine's experience with Humanyze brings some of this into focus.[47] This article offered an important perspective as one of the only pieces of public discourse written from the perspective of employees who wore sociometric badges for a two-week period. As Greg Lindsay summarized in the introduction to the employees' experiences with Humanyze, "Our goal was to discover who actually speaks to whom, and what these patterns suggest about the flow of information, and thus power, through the office."[48] Lindsay noted that while some employees consented to wear the badge, others did not. Fast Company did not allow Humanyze to be integrated into all communications; email and Slack were not part of the experiment. According to Lindsay's reporting, "staff members suffered the badges in silence," complaining about their "uncomfortable" weight and the need to wear the badge as a necklace rather than something "less intrusive." Others were concerned about the lack of explanation about how data collection worked: "Some found this Orwellian; others reported being lulled into complacency by its low-tech appearance and cheap plastic casing. Still more wanted feedback: Was this thing on? Was I doing this right?"[49]

At the end of the study period, Humanyze provided employees with a network map called the "true org chart," which organized groups based on their frequency of interaction. Lindsay suggested that "unearthing those relationships, understanding and visualizing them, is perhaps the most potent thing sociometric badges can do," but he seemed extremely ambivalent about what, exactly, the impact of this might be for smaller companies. As just one example, he noted how director of product Cliff Kuang quipped, upon seeing that seven digital writers talked more among themselves than workers in any other department in the company, "Surely, they must be complaining."[50] Lindsay suggested that Humanyze's reports on individual workers "didn't do much to improve" their understanding of their jobs, and another employee dismissively suggested the analysis seemed more geared to "productivity and efficiency" suggestions.[51] While acknowledging there is some utility, Lindsay also noted the badges themselves were hard to understand and inspired paranoia among those who wore them. The experiences with the Humanyze badge reported by *Fast Company* indicate that there is always some level of oppression in management science, forcing workers to become hypercognizant of their own behaviors or to fret about what their data might indicate about performance, productivity, and company culture. In this way, Humanyze makes "time-discipline" something closer to "data-discipline."[52]

ANALYZING THE MILLISECOND

Waber's investment in the millisecond as the ideal unit for monitoring employee behavior imagines a more pervasive version of management science. No longer is it just about discrete elements of job performance, like time spent talking to customers or time to complete administrative duties. Instead, the entire daily practice of one's job becomes subject to datafication, supposedly converting the everyday habits of a worker into standardizable measures. In his public talks, Waber has suggested, "Surveys and consultants are useful for particular purposes, but can't give the same granularity at daily, even monthly about what's going on."[53] In mapping all potential milliseconds of interaction, the sociometric badge supposedly exceeds the limitations of established social science methodology like surveys. Rather than giving employees the option to respond to survey prompts to access the social world of an office, Humanyze presents datafication as the pathway to what is objectively true in workplace operations. Data science, for Waber, trumped social science.

Through its emphasis on datafication, Humanyze also turns employees into test subjects. In the Bank of America service center study, the Humanyze team decided to use A/B testing, which tweaks behaviors by putting one group in a variable group and the other in a control group and measuring differences to understand behavior within a population. Waber has described in interviews how his company aggregated data that call center supervisors were then allowed to review: "Bank of America was able to see in their dashboard essentially that if people's groups got more tightly knit, the more tightly knit your group got, the more cohesive your group, the quicker you completed calls."[54] After giving employees more time to talk with one another during lunch, Humanyze claimed the following: "Network cohesiveness, which measures how well they communicate, went up 18 percent. This reduced stress (as measured by tone of voice) by 19 percent. All of this led to happier employees and lower turnover rates, which went down 28 percent. The key metric though, call completion time, improved by 23 percent. These are numbers that on a scale of Bank of America could translate into billions in savings."[55] It is important to note that of the five findings presented here, only two—call completion time and lower turnover rates—are easily measurable. To measure stress through tone of voice is to insist on affective computing, which critical technology scholars from Rosalind Picard to Blake Hallinan have insisted is a reductive evaluation of human emotion.[56] A measurement of "how well" communication occurred is also spurious. Did employees report their own happiness, or was this likewise inferred through some means of datafying affect? Finally, the claim that call completions going 23 percent faster "could translate into billions in savings" is also curious, as there is no explanation in this oft-told story about how the calls saved the company money.

For Waber, this study demonstrated some of his distrust of social scientific studies: "They use surveys, use human observers, but we all know how that's limited."[57] Consider how Waber, to supposedly correct the limitations of social scientific methods, discussed the data collected through the sociometric badge's microphone: "[The data are] not audio content. . . . It's essentially tone of voice, volume, how quickly you speak, location within the office. . . . And no one had ever looked at any data like this before. The question was, Well could we maybe predict performance? Long story short we could, really, really, accurately."[58] Focusing on tone of voice, volume, and speed codes speech in particular ways for datafication. While Waber dismissed the human elements of social science work, he conveniently overlooked that humans—including himself—had to build the algorithms through which they process this audio content, how it is coded, and other elements. This became, across his interviews, a kind of science without scientists, in which wearable sensors and algorithms can render supposedly more accurate understandings of what makes workers more productive.

In this model, the research assistant combing through interview transcriptions as part of an initial coding process is replaced with a sophisticated computer program capable of doing that work automatically. Waber downplayed (if he acknowledged at all) the work that went into building these programs, or any real discussion of how the algorithms and protocols were developed, trained, and refined. Such omissions are important. They reproduce an ideology David Beer has called "the data gaze."[59] For Beer, the concept of the "data gaze" summarizes the assumptions undergirding companies like Humanyze: chiefly, that manufacturing and analyzing quantified data will drive "better" understandings of human behavior through granular analysis. The proponents of this ideology are rarely reflexive (at least publicly) about the limitations of datafication, choosing instead to proselytize its value as a magical tool of discernment and knowledge production. Such advocates do not often acknowledge the role they and their colleagues play in building systems of extraction and analysis, preferring instead to treat machines as detached and objective observers of human behavior.

One of Waber's recurring lines about Humanyze is that "essentially you get a Fitbit for your career." As he went on to hypothesize: "Imagine, for example, you're a programmer, and you say, 'Well, I want to be the best programmer.' Well, we know in your company what the best programmers do, how they actually work, and we can show you what you do and how your compare to those people."[60] As in other interviews, there are many slippages here: How does one define "best," or "actual work," or even what measures are being compared? As chapter 3 explained, Fitbits and similar devices do allow for comparisons of behavior across a population. Likening Humanyze to Fitbit is another means of repackaging behaviorist dreams of using technologies to program human behavior through a vaguely defined sense of optimization. As a glowing profile of the company in *Computer World* imagined, "Programmable lighting and flexible office layouts, for example, could be automatically configured to 'nudge' employees in the organization to behave differently. . . . Email could be delivered more slowly when an individual's workload is high, and certain emails could be emphasized when communication with a particular coworker is overdue."[61] Again, workers are imagined as programmable, their habits able to be redefined and refined through suggestions and nudges.

Rosalie Chan, in a profile for *The Week*, also reproduced the belief that such surveillance measures create objective, value-neutral truth about what goes on in an office: "What humans report can be different from what's actually practice, but AI-collected data can provide the clear-cut truth by sifting through employees' emails and chats to collect data on how work is done."[62] For Chan, problems of unethical surveillance can be solved through fine-tuning regulations, rather than by questioning the system. Indeed, as science and technology scholars including Ruha Benjamin, Virginia Eubanks, and Winifred Poster have argued, many artificial intelligence systems are designed to exclude, marginalize, or otherwise discriminate against various groups of people who are flagged, sorted, and classified in stereotypical ways.[63] These public commentators, by contrast, have eagerly called automated data manufacturing and analysis a means for providing "the clear-cut truth." Humanyze is offered as documentation of daily life rather than a constructed representation. Such reporting promotes the intersection of surveillance and daily life, imagining it as a means of generating profits and efficiency while figuring workers as a series of legible datasets.[64]

Humanyze's focus on productivity and sociability does not dissociate it from earlier models of monitoring in industrial capitalism; if anything, it just rearticulates them through datafication technologies. Be it through the timecard, the stopwatch, the security camera, or the sensor-filled "sociometric badge," workers often exist at the mercy of data collection projects designed to rationalize their actions. As mentioned in the introduction to this book, *manage* originally meant to *bring to hand*. Through devices like Humanyze, wearables extend the work of management quite literally to the body: if not to the hand, then to the neck and chest, where the sociometric badge sits. Humanyze aims to bring management ever closer to a worker's body. In a 2015 profile of Humanyze, Ron Miller of *Tech Crunch* described the company's particular innovation: "Business to this point has not used data effectively to measure the productivity of its workers. They may look at spreadsheets about overall business performance, charts about sales conversion rates and the number of leads generated by marketing, but they don't look at how employees interact with one another and what impact that has on the bottom line."[65] The voice monitoring, locational tracking, and behavioral analysis Humanyze offers extend the disciplinary aspects of labor, measuring all forms of social interaction as part of the production of capital. As Melissa Gregg has suggested, "Productivity holds such profound power as a rationale for work because it manipulates an ethical dimension that lingers from the original spiritual foundations of capitalism."[66] Productivity, perhaps tautologically, produces; it is endemic to the work of capitalism, and it relies on time-discipline. While Waber has focused on the "millisecond" to position Humanyze as extending time-discipline, the company seems as much if not more interested in quantifying and making determinations about the social realm of workplaces, treating those supposedly ineffable moments of what was once called water cooler chat as a site for governance.

CONCLUSION: CULTURE ESCAPES

The goal of people analytics seems to be broadly to datafy and optimize as many elements of human behavior (and especially labor-related behavior) as possible. Though this ideology treats behavior as an inherently rational, analyzable, and potentially controllable system, Waber has occasionally acknowledged its limitations. As he has bluntly lamented: "There isn't a general understanding of what the term *culture* actually means quantitatively."[67] Raymond Williams famously called culture "one of the two or three" hardest words to offer a consistent and coherent definition for in his *Keywords* project, and Waber also confronted the ineffable components of culture as something that is fundamentally elusive of datafication.[68] In an interview with *Vice*, Waber explained that "face-to-face is much more predictive of the things we care about than digital data."[69] Again, this is a bit of misdirection: converting conversations into audio files on some level treats human behavior as data. As Waber and his associates used the sociometers to produce data on worker behavior, they themselves were not necessarily interested in observing face-to-face communicators. Datafication presumes face-to-face behaviors become legible and interpretable once they are rendered as data. Waber admitted in *People Analytics* that he had little interest in ethnography, seeing it as a biased form of social science that does not offer much in the way of objectivity. People analytics and the sociometric badge reinforce conceptions of datafication as a value-neutral and superior mode of understanding human labor. Waber delegated authority to his badge, evading discussions of how it was built and rarely acknowledging any biases the data may have.

Humanyze is just one of many datafying technologies—wearable or otherwise—that cause a collision of mathematics and culture. As Ted Striphas has explored in his sketches of algorithmic culture, there are increasingly practices, operations, and events that draw "what was long taken to be the conceptual *sine qua non* of qualitative human experience—*culture*—into the orbit of computational data processing."[70] The lived experiences of workers—the culture they build through their work and their negotiation of daily tasks and routines—escape the datafication of the workplace at least to some degree. And this is, largely, the point of this chapter's attention to how Waber has talked about Humanyze, and how many reporters in turn have reproduced his assertions about the product. To use a wearable device to analyze labor depends on granting authority to datafication, positioning an amalgamation of sensors as inherently more useful than interviews or observations. Such debates about quantitative and qualitative apprehensions of the world and fights over the correct meaning of *empirical* have long characterized academic research among many disciplines.[71]

While this chapter has explored the discursive claims of Humanyze's representatives and considered them alongside the genealogies of management science to which it is indebted, Waber's offhand remark about culture may be the most telling, and points to where this book will ultimately conclude: attempts at instituting new processes of administrative culture must simultaneously exist in tension with culture as a site of negotiation and resistance.[72] Culture is fundamentally irreducible; it is built from maps of relationships that do, certainly, include technology, but technology cannot determine culture outright.[73] Waber's lamentation would seem to come from this recognition that sensor-based datafication cannot render humans totally knowable and programmable.[74]

My goal in this chapter has been to demonstrate how some articulations extend conditions of surveillance. Datafication-based surveillance is offered as a strategy for solving supposed problems of workplace inefficiency and employee dissatisfaction. The focus in this chapter has been on Humanyze and its sociometric badge, but this case study has just been one pathway for considering how different institutions strive to know things "better," and how the decision to monitor human behavior is, as it has been since the development of management science, tied to arbitrary measures of productivity, efficiency, and time-discipline.[75] To continue building an understanding of how wearables are pitched as solutions, the following two chapters move on to concerns about how to build accountability for law enforcement through body-worn cameras (chapter 5) and how to improve the operations of space through near-field communication on wearable bands (chapter 6).

POSTSCRIPT: COVID-19 AND THE TRANSFORMATION OF WORK

The primary research for this chapter was done in the fall of 2019, before COVID-19 reshaped seemingly every layer of public life. One of the effects of the pandemic has been a transformation of workplace technologies as videoconferencing, remote work, and hybrid workplaces all became increasingly common even after the emergency phase of COVID-19 passed and vaccines became available. At the same time, employee dissatisfaction and turnover skyrocketed in many industries, with the catchphrase "The Great Resignation" referring to a tendency for employees to leave jobs and careers entirely during the pandemic.[76]

Humanyze likewise pivoted during the pandemic, publishing a series of articles on its website about how to navigate workplaces during remote work and the "return-to-work" period of the pandemic in 2021–2022. In 2020, for instance, Humanyze published a study that sought to help an unnamed company "maintain and improve their overall organizational health in the midst of this sudden disruption."[77] Here, Humanyze focused on what it calls "collaboration data," measured from "calendar," "email," "regional location," and "Slack (chat)." The case study did not actually discuss how the data was collected, which data was taken from email, or how it was analyzed, but Humanyze nevertheless calculated "Organizational Health Scores" indicating the company's health dropped from a 6.4 to a 5.8 during the transition to remote work (what this score means, and how it was calculated, were not discussed).[78]

During 2021, Waber pivoted more to this vaguely defined "organizational health" on Humanyze's blog.[79] Organizational health, according to Waber, "refers to its ability to unify around a common vision, function effectively, weather change, and use creative innovation to grow from within."[80] He argued that metrics related to engagement, productivity, and adaptability would all be important for maintaining and improving this organizational health during COVID-19. Using findings from the 2020 study, Waber began touting metrics such as "'Workday Span' (how long someone's workday is on average) and 'Manager Visibility' (calculated using meta-data from communication tools and channels)."[81] As hybrid work became more accepted in 2021–2022, Humanyze pivoted again to focus on employee retention.

While the company's website still touts its sociometric badge, it has become far more focused on how to "Drive Performance & Retention" and features strategies for human resources managers to adjust work schedules and gather data about how different employee groups interact. Implicitly, the badge became incompatible with ways COVID-19 transformed work. This led to the creation of the Humanyze Platform, "which measures anonymous corporate data against decades of MIT research and analysis of over 20 billion workplace interactions" to help organizations find "success." The company's August 2021 press release touting its workplace strategy solution for the Humanyze Platform made no mention of the sociometric badge, though it touted "data-driven benchmarks, indicators, and metrics" as well as "an award-winning, patented AI platform."[82] The wearable badges were somewhat obvious surveillance devices, their presence obvious to workers. The Humanyze Platform appears to be more diffuse, datafying work through software and aggregating it into an ever-growing mass of company data used to refine understandings of human behavior. Much like COVID-19 mitigation demanded community surveillance testing, so too has Humanyze used the pandemic to advocate for ongoing and pervasive surveillance of employee behavior as part of its own "organizational health" assessments. This postscript speaks, if nothing else, to the ongoing transformations of workplace surveillance technologies and how articulations between wearable technologies and labor shift alongside changes in other domains of life.

FIVE

Law Enforcement

OR: THE OPACITY OF BODY-WORN CAMERAS IN UPSTATE SOUTH CAROLINA

ON APRIL 4, 2015, police officer Michael Slager pulled over Walter Scott in North Charleston, South Carolina. Scott, who had an outstanding warrant for unpaid child support, fled. Slager pursued. There was an altercation. Scott broke free and ran. Slager shot Scott—an unarmed Black man—eight times in the back, killing him. In his report, Slager wrote that Scott attacked him and grabbed his Taser, necessitating the use of lethal force in self-defense.[1] A week later, cell-phone video surfaced from a nearby onlooker who had recorded the incident. The video showed—clearly and horribly—Scott turning to flee, and Slager taking a shooting stance, then shooting Scott until he fell and died. The video demonstrated Slager's report was false and led to community protests against police brutality in Charleston. The video of Scott's murder was hailed by some outlets as a demonstration of the importance of citizen journalism, the capacity of cell phones to provide some modicum of justice, and the ability of technology generally to increase knowledge about what exactly happens in events when someone is murdered and unable to defend themselves. For others, it demonstrated the possibility that sousveillance—or the monitoring of those in power—could become meaningful with the proliferation of video-recording devices.[2]

Scott's murder happened in the midst of a surge of activism in the United States around Black Lives Matter, an organization devoted to policy reforms addressing systemic racism, including police brutality.[3] On the heels of the high-profile deaths of Michael Brown in Ferguson, Missouri, Eric Garner in New York, New York, and Freddie Gray in Baltimore, Maryland (among, tragically, many others), national protests pushed for justice for these and other Black bodies as part of social and political debates about the role of members of law enforcement and their relationships to communities. While Slager was convicted of murder and sentenced to twenty years in prison, Scott's murder spurred other efforts at reform across the state.[4] It was the impetus for South Carolina state legislators to develop, pass, and implement guidelines for law enforcement use of wearable cameras to record (supposedly and ideally) every interaction an officer has with a citizen. The goals here were twofold. First, the cameras were imagined as a behavioral deterrent, whose mere presence would encourage officers and citizens to behave according to legal and social norms. Second, the cameras would create a comprehensive archive of officers' interactions with citizens. Through this archive, footage could be distributed to interested publics as evidence of good behavior, as well as to arbitrate high-profile or violent encounters. However, as this chapter demonstrates, the promises of transparency were routinely undercut by the technical, infrastructural, and legal apparatuses, which have only served to reinforce opacity around police videos.

The articulation of wearable technology to law enforcement, and the promise of more just (or at least more accountable) policing positions this chapter squarely within discourses of technological solutionism, or the belief that technologies themselves are agents of change. While everyone from then president Barack Obama to local police chiefs were clear to not call body cameras a "panacea" to police killings, body cameras were also repeatedly positioned as valuable tools for transparency and accountability in law enforcement.[5] This chapter charts how wearable cameras became articulated to the processes of law enforcement through a case study of body camera laws, their implementation, their governance, and the controversies surrounding them in Greenville, South Carolina, a small city in the northwestern part of the state approximately four hours from the larger Charleston area where Scott was murdered. It draws on reporting from local newspaper *The Greenville News*, as well as state legislation, local policies, police statements, and videos, to trace the tensions in mandating that police officers use body-worn cameras.

Though the state legislature and then governor Nikki Haley moved quickly to pass laws directing all police departments in the state to adopt body cameras to try to prevent further murders from occurring, these laws were just the beginning of a long process to acquire cameras and servers, develop protocols, train officers, and suitably record and store the videos. Even then, Greenville Police have been criticized by local lawyers, journalists, and activists for shielding violent altercations from public view or for neatly packaging them in "community briefings" that tell the police's side of the story without providing access to unedited footage. At the heart of this chapter are three different policies, which each impacted the relationship between law enforcement and wearable cameras: (1) the initial state law signed in 2015, directing police departments to devise wearable camera mandates specific for their departments; (2) the Greenville Police Department's mandate, which further refined guidelines for how to record and store interactions with civilians; and (3) the South Carolina Freedom of Information Act office, which mediated the ability to access wearable camera footage. In particular, the laws create a loophole exempting most body camera footage from Freedom of Information Act requests, which has continued to frustrate journalists trying to report on police encounters and accountability in the years since these body cameras were implemented.

In exploring the articulations between wearable cameras and law enforcement, this chapter charts a set of conflicts and struggles that are meant to explore the hard work of articulation in a single location. Wearable camera technology sits in relation to officers and departments, lawyers and judicial proceedings, citizens and activists who have a vested interest in learning more about policing in their community, and journalists who struggle to access information on incidents. Police departments obscure access to this footage. This process of bringing order has gradually reinforced, rather than challenged or made transparent, the operations of policing in Greenville. In focusing on South Carolina and the Greenville area, this chapter can clearly not account for the variations and complexities of wearable camera implementation. However, as the first US state to create a body-worn camera law, South Carolina serves as a useful space in which to examine how claims to transparency have only led to more concerns about opacity. C. J. Reynolds has called these practices "mischievous infrastructures," which allow police departments substantial control over what footage to release, when, and to whom.[6] Through triangulating the police departments themselves, the journalists working for the *Greenville News* trying to report on body-worn camera policy and violent policing incidents, and activists in organizations like Fighting Injustice Together, this chapter follows over a half decade of controversies and contestations over policing in the Upstate region of South Carolina, and how wearable cameras have been centered in discourses about law enforcement.

BODY-WORN CAMERAS AS TECHNOLOGICAL SOLUTION

Body-worn cameras were directly correlated to protests against police brutality in 2014 and 2015. Recorded material—convenience store surveillance footage, cell-phone videos, and police dashboard cameras—played important juridical functions in motivating protests.[7] The question of what recorded images revealed about the dead Black bodies and the officers who fatally shot, strangled, or otherwise killed them became the subject of outrage in local communities, on news channels, and across online networks, while also allowing respective police departments to claim nothing illegal had occurred in most of these instances once the cases went to court.[8] Police reform became a national issue during this time.

The implementation of body-worn cameras has occurred alongside a transformation in policing practices with so-called predictive policing. Sociologist Sarah Brayne studied the Los Angeles Police Department's incorporation of predictive policing.[9] As Brayne unpacked through how police forces license technology with companies like Palantir, officers have had to negotiate how to bring computational technologies into their workplace, and these technologies run the risk of reproducing biases in the data they receive, such as targeting predominantly Black neighborhoods because police were already overpolicing and making more arrests in those neighborhoods. The generation of more data, be it in algorithmically derived recommendations about resource allocation or in the increased production of recorded audio and video, has not necessarily challenged policing practices as some activists had hoped.[10]

Police departments implementing wearable cameras have done so explicitly or implicitly under the guise of transparency, suggesting these cameras will be able to curb problems of police brutality, or at least ensure that when a citizen is killed by the police, a record will exist as legal evidence of whether the officer was justified. In the wake of the massive Ferguson protests in 2014, then president Barack Obama formally requested over $200 million for the purchase of and training in use of body cameras for police officers in the United States.[11] In 2015 Obama's Justice Department announced a body-worn camera pilot program that would provide $19 million to purchase body-worn cameras, $2 million for training and technical assistance, and $1.9 million to examine the impact of their use.[12] The program was part of Obama's commitment to provide law enforcement agencies with fifty thousand cameras by the end of 2018.[13]

At the time, Obama cautioned: "There's been a lot of talk about body cameras as a silver bullet or a solution. I think . . . there is a role for technology to play in building additional trust and accountability."[14] Also in 2015, Hillary Clinton called giving every police department in America "a common-sense step" for promoting, again, transparency and accountability.[15] In these and other proclamations, the camera was supposed to provide access to the truth of what happened. These transparency claims were, from the start, deeply at odds with what scholars of still and moving images have found for decades regarding how recordings actually construct and restrict understandings of "what really happened."[16] In the years since 2015 these claims have generated a wide-ranging debate regarding technological solutionism, in which the technology of wearable cameras is framed as the way to address more systemic problems of training, practice, litigation, and indeed, racism.[17] There are at least three major themes that show up in reporting and research: the opacity of body camera footage (if it is even recorded at all),[18] the reliance on police departments to package and distribute the footage,[19] and concerns about whether or not these devices have actually created empirically measurable improvements in policing practices and community relationships.[20]

This solutionist conception of body cameras makes several assumptions about how they work: first, it assumes the camera is always on and always capable of capturing an event perfectly from the chosen perspective. As Emmeline Taylor has demonstrated in Australia, a number of police cameras have to be turned on by the officer, and officers have discretion over when their documentation would be a violation of citizen privacy, giving them the authority to turn off their cameras.[21] Stacey E. Wood has further argued that digitally storing police body camera files in cloud-based archiving systems involves a host of complex legal questions about how these documents might be admitted as "authentic" evidence: "The complexity of technological systems, the distributed nature of documentary forms, and the various levels of technological expertise of involved parties leave traditional methods of assessment and determination wanting."[22] Further, it is often unclear how this archived material is treated as a matter of public record, and whether or not it will be available to the public.[23] Some officers have discretion over when their cameras can be turned on and off, and many of these files are stored in complex cloud-based archiving systems that make responding in a timely way to FOIA requests an ongoing problem. Indeed, critiques of the cost and ineffectiveness of wearable cameras became one of the few issues uniting ideologically disparate outlets such as *Vox* and *Fox News* in the late 2010s.[24]

A *Washington Post* op-ed from 2018, titled "The Ongoing Problem of Conveniently Malfunctioning Police Cameras," alleged that officers only record when there can be no reasonable debate about the appropriateness of their actions.[25] Problems with treating wearable cameras as components of accountability extend to the security of the devices themselves. In August 2018, *Wired* reported that at least five major body cameras, including those from Vievu—the company that manufactures the New York Police Department's cameras—had "security issues that could allow an attacker to track their location or manipulate the software they run."[26] Some platforms had unsecure software or cloud storage, meaning the archived videos of officer action were left vulnerable to hacking, editing, and manipulation. Research has been ambiguous at best about the actual efficacy of cameras in reducing violence among police officers and improving community relations. In October 2017, a Washington, D.C.–based study followed one thousand police officers who were randomly assigned body cameras for a seven-month period, finding that "the effects were too small to be statistically significant. Officers with cameras used force and faced civilian complaints at about the same rates as officers without cameras."[27]

The Greenville Police used the Panasonic Arbitrator body-worn camera.[28] Like most body-worn cameras, it is a small, square, black camera meant to be affixed to a wearer's chest. The camera measures about 3.4 by 2.75 inches. The top third has a camera lens, with a microphone recessed into the camera to prevent environmental noise like wind from interfering with recording human voices. A button on the bottom half of the camera begins the recording process. Videos cannot be played back on the camera itself, and the Arbitrator camera must be connected to the dedicated software, the Arbitrator Evidence Management Software solution (or SafeServ), to store its footage. This is designed to help secure footage and prevent it from being manipulated in-camera.[29] While many officers wear the Arbitrator on the chest, it can be worn as a head mount, chest mount, or belt mount, providing some flexibility in where it is located or allowing departments to mandate wear in some areas to better assist with video capture. While the camera does not have a playback screen, there are several buttons to begin recording, emphasizing the act of beginning a recording as the most important feature.[30]

The design of body-worn cameras is relatively standardized. Most of them are small, rectangular, and black. Axon, another leading company in this space, also places its camera lens in the top third and a large activation button in the bottom third. Some more recent models include what Axon calls "flex" cameras, which are connected to the main body-worn camera via a wire and provide a second or third camera that can be attached to a shoulder or hat, or worn over the ear.[31] Like Panasonic's Arbitrator, these cameras do not have many controls for officers in the field beyond the button that begins recording. Like Arbitrator's dedicated system for storing videos, Axon has developed the Axon Respond+ communication platform, "equipping wearers with the ability to communicate with command staff and specialists in real time."[32] This offers police more control over video and supposedly allows for coordination between different members of the department to help de-escalate intense confrontations. Unlike an audio-only police radio, the Respond+ platform integrates video to help law enforcement officers process situations and recommendations in real time. While the external appearance of the camera hardware changes relatively little across companies, Panasonic and Axon—as just two companies in this industry—have developed features that are not about transparency and accountability to the public so much as they are meant to help officers record and store evidence on proprietary platforms, on the one hand, and communicate with teams in real time, on the other hand.

This sketch of body-worn cameras demonstrates the institutional challenges of actually articulating these devices to the practice of law enforcement and policing and how the devices are routinely imagined to do more than they may very well be capable of doing. From here, I turn to the case study of this chapter, which is divided into five subsections reflecting chronological developments in the implementation of body-worn cameras in Greenville, South Carolina. Most of this research was performed through an analysis of public reporting and policy documents. The *Greenville News* and its online archive of reporting from 2014 to 2023 was an instrumental asset in finding records of body camera footage released to the public, links to police documents, interviews with police chiefs and sheriffs, and reactions and comments from local activists. These public debates and policy documents indicate that claims about bringing order often depend on overstating technologies' capacities to solve public problems and creating a set of policies that prevent solutions from manifesting.

SOUTH CAROLINA: DEVELOPING AND PASSING THE POLICY

As discussed at the beginning of this chapter, Walter Scott's murder became the catalyst for politicians to unite in bipartisan support of wearable camera policy. While Scott's murder expedited and framed the use of police body cameras as a potential civic good, it does not provide the whole story about the difficult implementation of body cameras in Greenville and the surrounding area. Greenville County contains about five hundred thousand people, with about seventy thousand living in Greenville City. From the 1980s through the 2000s, the city underwent a relatively publicized urban renewal project designed to invigorate the economy of the region and boost its larger reputation in the state and the southeast.[33] In December 2014 the *Greenville News* reported that many law enforcement agencies in the region had already experimented with body cameras, providing them to some officers to test their efficacy and feasibility before purchasing cameras for an entire department or agency. Notably, the *Greenville News* pointed to the costs of purchasing and updating cameras, as well as storage space. Local officers in nearby Pickens County discussed the difficulty of establishing standards for how often officers should turn on cameras and how long recordings should be stored; in other words, to what degree it is useful to capture and store the everyday lives of the police officers who wear the cameras.[34]

This process of standardization was taken up in the state legislation. In early 2015 state legislators proposed a bill mandating "all law officers in South Carolina to wear cameras that would record all contacts they have with the public. Officers would have to tell people they are wearing the devices, and everything recorded would be retained under existing policies governing law agencies."[35] The Walter Scott shooting occurred as this bill was being debated. An editorial by a resident of Simpsonville, a town in Greenville County, urged lawmakers and law enforcement officers to push for cameras: "A comparison of news reports about the incident published before and after the video was released shows what an impact indisputable video evidence can have on our perception of events. . . . The simple fact is that if all police officers are required to wear video cameras operating completely out of their control, we do not need have to worry about incidents like these going unpunished."[36] This passionate plea is nevertheless a fantasy about fixing problems through technology. It continues: "The knowledge that their actions are being recorded forces officers to always behave in a professional and legal manner. The potential for increasingly accurate evidence collection and improved police behavior make modernizing the police force an easy choice."[37] Here, the letter writer is a technological determinist placing faith in the technology itself to automatically alter human behavior. As discussed in the previous chapter, the assumption is that technologies can change behavior.

Some police departments did not necessarily share this sentiment. In nearby Greer, South Carolina, police chief Dan Reynolds did not want a state mandate to dictate body camera policies: "I don't think the camera issue is resolved yet. We're still learning how to use these things."[38] At the time, Greer police officers were already wearing cameras: "Recordings begin as soon as officers respond to calls for service, and they carry on throughout interaction with the public. They are kept for 60 to 90 days, longer if they have value as evidence in investigations or for training."[39] Additionally, cameras could not be turned on in private residences unless residents gave consent, and only one officer had access to edit the footage (an attempt to prevent corruption and manipulation of video evidence). At the time, Victoria Middleton, who served as executive director of the South Carolina branch of the American Civil Liberties Union, agreed with the need for a balanced approach to camera policy: "While they can be helpful and we're supportive, generally, they're not going to be a foolproof solution to the problem of law enforcement engaging with citizens. They need to be used properly, and oversight needs to be in place. They can be very, very valuable, but they can also be intrusive just like any recording devices."[40] Here, several tensions are on display: local departments want the freedom to develop their own policies in accordance with their communities. Chief Reynolds was not dismissive of body cameras; rather, he wanted policy to emerge from testing rather than from legislation. Middleton, as well, noted the potential importance of the cameras while imploring that policies be developed with substantial oversight to ensure citizens are also protected from additional surveillance.

Michael Nunn, who served as the department spokesperson for the nearby Florence County Police Department, also raised privacy concerns, but did so from concerns about how videos would be subject to FOIA requests. According to a summary of Nunn's testimony against the development of the state mandate, "people could be videoed whenever officers came to their homes for any reason, including criminal domestic violence," and "all captured material would be subject to the Freedom of Information Act."[41] Importantly, these concerns would be addressed in the final policy and its challenges in subsequent years, such that most body camera footage could actually not be accessed through FOIA requests, severely limiting their public reach. This is discussed at greater length later.

Despite these tensions, by April 23, 2015, a *Greenville News* editorial described a "national consensus" that had emerged around the decision to give police body cameras. As the editorial authors suggested, "The technology is available to remove most, if not all, doubt about what happens during encounters where police officers shoot unarmed people. The technology protects law enforcement officers from false allegations and perhaps even from their own impulsive actions. Body cameras also protect the public, first by not denying a dead person the right to have his or her story told in full, and second by allowing the community to see, rather than speculate on, what happened during a deadly incident."[42] The editorial affirmed a particular attitude towards cameras: they "remove most, if not all, doubt"; they protect officers; and they protect the public. They are framed as a win-win-win, a device that can serve to resolve the tensions in the community and in the nation. This is an example of a sociotechnical imagination, an instance in which the articulations between technology and society are imagined to be perfectly aligned and without significant complication, such that the technology becomes an obvious solution to a problem. Even though wearable camera proponents had been sure to caution that they would not be a magic solution, editorials like these demonstrate how those limitations are often downplayed. The important thing, so the argument went, was to get the cameras and make them work.

In mid-June, Governor Nikki Haley signed the bill into law, requiring "police agencies to create a policy on which officers will wear the cameras, when they should and should not be recording and how the videos are stored."[43] This created a mandate quite different from those discussed in the previous two chapters; instead of universities and workplaces being the institutions asking populations to don wearable technologies and produce reviewable data, here it is the state government. At roughly the same time, US senator Tim Scott was working on getting similar sorts of programs passed in the US Senate. In promoting his bill, Scott said cameras "bring clarity to very often contentious and unclear situations,"[44] again assuming that videos will be clear, available, and useful. South Carolina was the "first state to require the cameras be worn by law enforcement," and many supporters of the plan proudly pointed to this as a sign that South Carolina was actively working to use technology to solve problems.[45] However, initial restrictions on "public access to the recordings" sounded alarms for journalists and advocates, who worried the law would negatively impact their ability to assess altercations and incidents with law enforcement.[46]

In December 2015 the state's Law Enforcement Training Council approved statewide guidelines for body camera use. "Under the guidelines, all officers, uniformed or not, who regularly answer calls and interact with the public must wear the cameras."[47] The guidelines also provided rules for activation, which should occur "when an officer arrives at a call for service or initiates any other law enforcement or investigative encounter with a member of the public," which includes everything from arrests to "an adversarial contact or a potentially adversarial contact," and they should not be used during breaks or during any sort of personal time.[48] The policy also established guidelines for storing these recordings. All "non-investigative" recordings and those that do not involve an arrest should be deleted after fourteen days. More importantly, "the footage is not a public record subject to the state Freedom of Information Act, under the new law, although a law enforcement agency, SLED, or prosecutors may release recordings."[49] This decision created inaccessible records. While "those who are the subject of a recording," appropriate attorneys, and others can request "data from the recordings," this makes it difficult for journalists and general investigators to access and use these recordings.

The decision to not regard camera footage as part of the public record formed the crux of most criticisms of the body-worn cameras over the coming years. This policy decision effectively changed the stated aim of the cameras from being about providing understanding and accountability, making them instead part of an elaborate and costly storage operation. The footage appeared to no longer exist in the interests of the public; it exists in the interest of the police, who can decide to shield the video under most circumstances or package it in ways that promote their side of the story, such as the propagation of community incident briefings.

Once the policy had been established, it became up to the police departments to both adapt its guidelines for their own officers and sell the policy to their communities. Although the use of body cameras was a state mandate, in other words, local departments had agency to create some guidelines that best suited their officers. Later in December, Greenville City's police chief, Ken Miller, engaged in a public relations tour to discuss the policies with citizens and hear feedback from them. At an event at the Nicholtown Community Center, he described cameras "as an independent witness."[50] To describe the camera as "independent" is a fundamental misunderstanding of the camera. Because it is literally tethered to the officers, it is wholly *dependent* on their bodies. In treating the camera as a separate agent, Miller overlooked the articulation of body and device necessary for them to function.

In January 2016 the Greenville Police Department finalized its general order on body-worn cameras based on state guidelines. The policy provides guidelines for training, deployment, recording, restrictions, data collection, supervisor responsibilities, auditing, and retention and viewing. These protocols direct a proper set of rules around the habitual use of body-worn cameras, including directives on how and when to turn the cameras on when interacting with civilians, how to store the footage, and how to use videos as evidence.[51] In short, by the start of 2016, law enforcement agencies had developed plans to use the cameras in accordance with the lofty goals of the bill to improve transparency and accountability. However, as the next section demonstrates, implementing these policies was extremely difficult.

IMPLEMENTATION: CAMERAS IN PRACTICE

One year after the bill was passed, the *Greenville News* reported that little progress had been made on acquiring and implementing the cameras. The sheriff's office in Greenville County had "applied for grants and expected to receive funding this summer" to purchase cameras.[52] The Greenville Police Department and others had pursued buying their own cameras over time with available funds and were now seeking reimbursement for these purchases. Bruce Wilson, an activist who founded the Greenville-based Fighting Injustice Together, complained the agencies "are trying to focus on the dollars, even though they have the money in their own budgets. . . . It's a necessity that all deputies have body cameras. It helps reduce false complaints and it helps validate complaints. So why not have body cameras?"[53] Wilson's position acknowledges that body cameras may not solve the problems, but his choice of "reduce" and "help"—along with calling them a "necessity"—invests these devices with a political importance that suggests their articulation to law enforcement is required for the law to function properly. A month later, Greenville police chief Ken Miller outlined a five-year plan to improve policing in Greenville that included "$1 million over the next five years to outfit officers with body cameras using the latest technology," with funds coming from the city's budget.[54] Another way to say this is that the first year of the body camera laws was mostly characterized by a lack of body cameras. Although the state legislature and governor had quickly signed the policy, these delays in acquiring and implementing body cameras across the state serve as a useful reminder that articulation—even when mandated through law—takes time to come into effect.

In the first eight months of 2016, "South Carolina law enforcement officers [had] been involved in 30 shootings. Of the 18 cases for which South Carolina Law Enforcement Division records are available, only two of the shootings were caught on body cameras."[55] The bill and the policy guidelines establish rather clear protocol about how the technology should be used in relation to the everyday lives of officers, but these legislative and regulatory documents do not necessarily matter when departments are slow to acquire and implement the cameras due to the economic cost and the lack of existing infrastructure (or because, as Wilson argued, they simply want to avoid doing it for as long as possible). As late as August 2016, the Greenville County Sheriff's Office was still using VHS dash cameras to record police activity: "The old ones have been repaired so many times that most of them, or many of them, are beyond repair."[56] While body cameras may be framed as powerful tools, departments like the Greenville County Sheriff's Office point to existing problems with recording infrastructure as a reason to remain skeptical about any technology providing a quick solution.[57]

By October 2016 the state granted $135,000 to the Greenville County Sheriff's Office to purchase and implement body cameras. As the Sheriff's Office spokesman, Ryan Flood, noted in response to the plan to use the funds to purchase 125 cameras, "State law mandates that those uniformed officers whose primary function is dealing with the public be outfitted with them. So, we will get the 125, assess it and go from there."[58] The parameters of that assessment and how the office planned to measure the efficacy of the cameras were not explained. Even though the Sheriff's Office admitted its existing recording infrastructure was outdated and in need of constant maintenance, it defaulted to state law as the reason for purchasing the cameras in the first place. While the mandate created the need to articulate wearable technology and law enforcement, in other words, some departments and offices were framing this implementation as an encumbrance rather than something useful and desirable.

It took until March 2017 for the Greenville County Sheriff's Office to acquire the cameras and begin using them, with other law enforcement officers in Greenville beginning their use in the first week of May. Again, deference to the mandate as the reason for wearing the cameras continued from the spokespeople in the Sheriff's Office. In a March 14, 2017, discussion of camera implementation, spokesman Drew Pinciaro said: "Deputies' interactions will be recorded which will promote trust from the community in what we do. . . . Roughly 300 deputies will be equipped, which covers the state law that mandates deputies in uniform, who have the expectation of interacting with the public on calls for service."[59] Journalist Tesalon Felician described the cameras as a "necessary tool," and state senator Karl Allen claimed: "A camera recording is worth 1,000 words. It was necessary to help the community as well as law enforcement to eliminate discrepancies in the stories on both sides of the issue. . . . It's a win-win for both sides."[60] Activist Bruce Wilson similarly hailed the launch of the body cameras as "a good thing."[61] Despite having no actual empirical data on the efficacy of camera use in changing police behavior and establishing trust in the community, the alignment between journalists, activists, politicians, and (even if begrudgingly) law enforcement speaks to the articulation occurring around wearable cameras, which appeared to unite these different domains of media, politics, and law enforcement. Framing them as "necessary" and "win-win" speaks to the belief that wearable cameras would be capable of producing behavioral changes in both officers and civilians.[62]

As Ken Miller noted in a promotion for the launch of the cameras: "While they may not fully capture every event or every angle, these cameras are extraordinarily useful in evaluating interactions between the police and public, and in improving professional performance."[63] Whereas previous editorializing had treated the cameras as agents that would change policing, Miller's language here is hedging yet positive, preparing the public for a potential lack of change while still insisting on the overall benefits of the cameras. The *Greenville News* also turned to more voices not associated with law enforcement. Dari Seabrook, a member of Greenville Black Lives Matter, framed the cameras as "just another tool for surveillance used by law enforcement, that ironically has been used as evidence that the system is criminal, violent, and broken. A lot of the city's policies . . . seem discretionary and that's a huge concern for us,"[64] such as protecting videos from FOIA requests. Because video content is collected from all police interactions but only selectively released to the public, the cameras perform more of a surveillant function. They may be attached to an officer, but the lens points outward at civilians, capturing their faces and behavior.[65] In another interview with the *Greenville News*, Bruce Wilson reiterated his claims about the behavioralist operations of the cameras, claiming, "It curbs the attitudes of a potential suspect and officer just knowing the body camera is engaged."[66]

The implementation of the body-worn cameras had other effects on how the police departments managed recording technologies. As the *Greenville News* reported near the end of June 2018, the Greenville Police Department had "nearly phased out" the use of dashboard cameras, opting instead to treat wearable cameras as the primary source of recording. However, because the wearable cameras were "not subject to disclosure" under FOIA laws, this shift in recording created the real risk of even less public understanding of police behavior.[67] The executive director of the South Carolina Press Association, Bill Rogers, put it this way: "It's really problematic for public oversight of what in the hell police are doing." Police chief Ken Miller insisted "there's a reason to be protective of [police]," while also acknowledging the need for transparency to maintain public confidence.[68]

In November 2018 the Greenville County Sheriff's Office, under increased scrutiny to provide timely public release of camera footage, introduced a new measure of transparency. In the case of circumstances such as deputy-involved shootings, Greenville's police department planned to create "critical incident community briefings," video presentations consisting of bodycam footage, dashcam footage, 911 calls, and other records that depict events, which were uploaded to YouTube complete with narration and explainers to help viewers understand the context.[69] This is a protocol borrowed from the Los Angeles Police Department, which had been creating its own community briefings since April 2018. These highly manufactured and produced packages claim that they do not draw conclusions, but these videos are quite obviously designed to present the LAPD's side of the story. They are carefully edited, with authority figures guiding and crafting the narrative. The officers and citizens involved are removed, leaving only the institutional and public relations voice of the LAPD to provide context.[70]

Such community briefings have much in common with what surveillance scholar Daniel Grinberg has called "transparency optics." According to Grinberg, official agencies often release documents designed to satisfy members of the public while still withholding information; transparency is a negotiated process, in other words.[71] In March 2019 the *Greenville News* reported that no such briefing was provided in the designated time window "after the Sheriff's Office's first deputy-involved shooting under the new program" because the person involved was a juvenile.[72] When the Sheriff's Office decided not to release a community incident briefing, its representatives noted that the person was both a juvenile and was not actually shot; officers fired at him but did not hit him. As the *Greenville News* pointed out, "Whether the teen was struck by the deputy's bullet should not have impacted the release of body camera footage in the briefing, based on previous comments from the Sheriff's Office."[73] Captain Tim Brown's response to questions about why the video was not blurred or edited so as to remove the boy's identity simply suggested, "We can't put ourselves in any kind of liability."[74] While there are very real concerns about the interactions of young people with the police, this incident underscores how the Sheriff's Office moved its own goalposts from a position of absolute to relative transparency, asserting the need to evaluate whether or not to create a community incident briefing on a case-by-case basis.

Later that same month, the police department released its first community briefing in response to a shooting.[75] The community incident briefing is an example of "deathlogging," or the ability to use mobile and wearable cameras to record the moment of human death, then use that footage to reconstruct and analyze incidents.[76] While the police insist that these community briefings do not interpret the events or render judgment, the very act of editing and juxtaposition works to embed particular understandings into the public understanding of this footage. It is a form of selective framing giving police the authority to narrativize events.

Greenville's first critical incident community briefing was uploaded to its YouTube channel on March 25, 2019.[77] It begins with an introduction from Lieutenant Ryan Flood of the Greenville County Sheriff's Office Public Information Office. Flood says the viewer will see "relevant video and photographs" to provide "an accurate overview" of the shooting. Flood states the briefing does not draw conclusions about the police's behavior. The briefing cuts to Tim Brown of the Office of Professional Standards, who provides a background summary of the incident. His vocal summary sets up the 911 call, which is then played in its entirety, before introducing and playing the body camera footage. The body camera footage serves as supporting evidence for Brown's narrative of what happened. The incident entails officers responding to a call from a man who claimed he was being poisoned and bullied; when they arrive at his residence, they hear a gunshot and take cover behind an adjacent residence. A woman, claiming she's been shot, exits the residence, followed by the man. The footage continuously freezes to isolate the suspect in the background, and Brown occasionally provides voice-over to narrate the officers' actions. The man does not comply with repeated requests to stop moving and get on the ground; he fires multiple shots into the air. Brown then says the man turned his gun toward the deputies, at which time they shot him. The video becomes blurry and unfocused when the officers fire their guns; the critical incident briefing does not actually show the man turning his gun toward the officers, and the body camera footage does not record this crucial moment. The briefing then cuts back to one of the police dashcams, which *does* capture the moment the suspect points his gun toward the officers. The briefing replays this moment in slow motion, ostensibly to justify the officers' need to shoot and kill the man. While the briefing claims to not "draw conclusions," the arrangement of footage and Brown's narration work to justify the officers' actions. The man is repeatedly heard asking officers to kill him, and he fires his gun in the air multiple times to provoke them. The officers' response is to continue to shout at him, which only escalates the situation toward more gunfire. There is no information about the man's mental health, whether an autopsy had been performed to indicate he had been poisoned or consumed some sort of narcotic, or any other information to help explain the incident. The role of the body cameras is to narrativize and implicitly justify police violence.

In June 2019 the issue of public access to these videos again underscored tense relationships between opacity and transparency regarding body camera footage, due to a new order from circuit court judges in Greenville County providing additional restrictions for defense attorneys; the supposed access to the everyday work of police officers was more a gesture to an ideal than something that was carried out in practice. A June 2019 order stated: "Defense attorneys have a right to receive copies of body-worn camera footage as it pertains to a case but that the footage cannot be shared with the public. If an attorney is found to be in violation of the order, they will be subject to sanctions including being found in contempt."[78] This contempt could include jail time. According to the judge who issued the ruling, distributing videos to the public "creates a substantial risk of violating the privacy rights of victims, witnesses and members of the public who may appear on such recordings simply as a result of interacting with law enforcement."[79] Whereas Black Lives Matter activists critiqued the surveillant operations of body cameras, this judge responded not by curtailing the use of cameras, but by curtailing access to the footage. The surveillance would continue, just farther from public accountability. As stated previously, in South Carolina, the FOIA allows law enforcement to withhold body camera footage, "making it difficult for the public to access unedited video of incidents involving law enforcement."[80]

At the end of June, this order was overturned by State Supreme Court chief justice Donald Beatty, citing that it "was counter to General Assembly's intent to promote transparency and accountability of law enforcement when laws were passed to require body camera use."[81] Both lawyers and journalists celebrated the decision, with SC Press Association attorney Taylor Smith saying, "It is particularly important that law enforcement not be able to shield their recorded actions with the public under specious arguments that releasing the nature of those interactions would somehow always violate a citizen's right to privacy."[82] Private defense attorney Frank Eppes noted, "The public paid for these cameras, they should have a right to view the footage."[83] Again, policy and judicial order manage what is done with the data wearable cameras collect. Transparency is no longer a given with body-worn cameras, but instead something to be negotiated with both police departments and judges.

THE PARADOX OF BODY CAMERAS

At the end of September 2019 the *Greenville News* began publishing a massive, multipart investigative report on the police departments and other law enforcement agencies in the region, detailing a decade's worth of law enforcement shootings, officer training programs, and more.[84] The series, called "Lethal Force," provided a comprehensive overview of law enforcement's failings to improve policing throughout the 2010s. As the *Greenville News* stated quite clearly: "Paradoxically, when South Carolina legislators passed a law in 2015 to require law enforcement agencies to use body cameras, the state moved a step away from transparency."[85] The language here is striking: what only four years earlier had been a massive rallying cry about the redemptive possibility of technology had now become a criticism of the efficacy of these same devices to fulfill their initial promise. The technologies, it seemed, had run into a problem: officers themselves. As the paper reported in its comprehensive review of officer-involved shootings, there were patterns of "neglect in body camera use" in the state's five largest law enforcement agencies: "For one out of every three to four shootings on average, there was at least one officer not recording."[86]

In May 2020 the murder of George Floyd in Minneapolis, Minnesota, set off a wave of Black Lives Matter protests across the United States, with renewed calls to reform or abolish the police. Protests in support of Black Lives Matter took place across Upstate South Carolina, including in Greenville and on Clemson University's campus. Much of the conversation in South Carolina refocused on the body cameras, which continued to be positioned as a commonsense means of improving policing and reducing violence. As SC Press Association executive director Bill Rogers put it in early August: "The time is right. People are behind this. Why would you have police video and not let the public see it? . . . The fact is that any time these cases have been brought to light, if it wasn't for that bystander video, it would have been passed off as defensible. I think it's the linchpin. It has to happen for police accountability."[87] Rogers's comments here largely ignored the previous five years. Police body cameras are *not* bystander videos; they are part of the ways police produce meaning about events, as with the community briefings that edit together multiple videos. The belief that bringing to light recorded video would, in and of itself, create meaningful reform in police activity seems almost foolish considering the previous half decade of evidence that these technologies could not act as the "linchpin" to accountability and reform.

The *Greenville News* did additional reporting on body camera efficacy during the summer, including a review of every certified law enforcement agency in South Carolina. The journalists found that "23 of the 178 agencies that were reached do not have body cameras," and that there were "at least 1,377 law enforcement officers who are not assigned bodycams though they work in agencies that have some bodycams."[88] As the report pointed out, despite the mandate from the initial 2015 bill, "The law leaves a loophole dropping the bodycam requirement unless agencies' programs are 100% funded, whether by the state or otherwise—and many are not."[89] Be it because of officer misuse of cameras, policies that prohibit footage from being disseminated, or economic conditions preventing police departments from purchasing cameras in the first place, the actual process of articulating wearable cameras to law enforcement in Greenville was largely failing.

One year after Floyd's murder, Bruce Wilson was still holding Black Lives Matter rallies and still pushing for bodycam footage to be made public: "We are tired of marching," Wilson lamented, before adding that "there ha[d] not been a meaningful change" in the year since Floyd's murder. Wilson continued to support use of body cameras and demanded footage be made available: "It is not an investigative tool. . . . Body camera footage was installed and enacted to create transparency for the community and law enforcement. . . . What good is it if only law enforcement can see it? It needs to be part of the FOIA request."[90] As of early 2024, the FOIA laws had not changed.

CONCLUSION: POLICING THE CAMERAS

In tracing the development and implementation of South Carolina's body camera mandate policy, as well as the subsequent negotiations, frustrations, and contradictions of that policy, I have tried to demonstrate the complexity of articulation in one local space. While it is tempting to read this chapter as a story of how wearable cameras fail to create the transparency many promised, it is more about the fallacy of investing in a wearable technology to do the work of transparency. While wearable cameras were promoted as bringing transparency, accountability, and order to policing practices, these claims almost always occurred in a vacuum, devoid of considerations about how policy, economics, and workplace cultures would shape the development of camera use. It is not so much that they achieved or failed to achieve their goal of providing transparency about the everyday practices of police officers, but that these goals were drawn from a belief that wearable technologies are able in and of themselves to solve civic and political problems. For frustrated community members, including Black Lives Matter activists, these cameras became an expensive means of police surveillance in communities. For journalists, accessibility became the primary problem: while the agencies spent money to download and store these videos, they were largely kept away from public view or else packaged and contextualized via law enforcement representatives.

While wearable cameras were embraced as a progressive solution following Walter Scott's murder, and South Carolina proudly touted its action in establishing statewide policies about body-worn cameras for law enforcement, the ensuing years have shown how the articulation of technology to law enforcement is fraught with debates about what these technologies should do, for whom, and when. The importance of the video showing Scott's murder was partly that it came from a citizen who happened to use their cell phone to record the violent crime. The release of footage from body-worn cameras all too often appeared designed to serve the police themselves, not the citizens who expressed a desire to review recorded materials. Although institutions like the Greenville Police Department use body-worn cameras in the name of accountability, transparency, and civility, these troves of visual and audio recordings are always opaque and inaccessible until circumstances force their release, often as community incident briefings. This outcome was, likely, inevitable. Ben Brucato has argued that positioning transparency as a precondition of accountability overlooks how video footage is part of how policing institutions protect themselves more than citizens: "The legal system privileges the police perspective."[91] Despite the early mobilization of accountability, the policies built around wearable cameras in South Carolina reinforced this privilege at the expense of providing access to concerned community members, journalists, and even families.

In mapping wearable technologies and their articulations to other domains, this chapter has demonstrated how this attempt to bring order to policing and community relations was paradoxical at best: invoking transparency while enacting opacity. This chapter has moved the book's analysis into public and civic concerns, demonstrating how wearable technologies become articulated to promised solutions and indicating some of the dangers of treating technologies as generally capable of solving these problems. In this chapter, the order that wearables promise to bring is no longer confined to one's body or one's workplace but extends out into political debates and public policy. The final chapter continues to consider how wearables are asked to work in spaces through a more sustained focus on their relationship to infrastructure.

SIX

Infrastructure

OR: THE DATAFICATION OF DISNEY WORLD

"JUST IMAGINE," implores the female-sounding narrator in a pleasant, upbeat tone. "Imagine entering your resort room, the parks, even making purchases, all from your all-in-one Magic Band." With its call to "tap into the magic" of the Walt Disney World theme park and resort experience, this promotional advertisement for the MyMagic+ technical system emphasizes independence, flexibility, and of course, wonder. Throughout the commercial, families are shown navigating Disney's theme parks—accessing rides, purchasing souvenirs, eating and drinking, entering resort rooms—all through the "magic" of the wearable MagicBand, a plastic identification band that datafies many of the daily operations of Disney World theme parks, including ride access, concession purchases, and event and dining reservations.[1] The commercial even emphasizes the magical wonder of this technology through incorporating what might best be described as a "fairy twinkle" sound effect (like one might associate with Tinkerbell or other characters who appear in the Disney parks) every time a Magic Band is shown in close-up touching a sensor station.

This final chapter pivots to a different set of articulating relationships between wearables and infrastructure. Infrastructure encompasses an array of material and social practices. These can be material (like bridges and roads) or metaphorical (like support services and emergency response systems). Infrastructure is also, conceptually, a structuring force. It helps to give shape to the world. In a very real sense, this book has been building an understanding of wearables as a structuring force in relationships between technology and multiple domains of human life. In this final story about articulation, this chapter looks at how the infrastructure of a space like Walt Disney World transformed to accommodate the Magic Band, and how the wearable assisted the park managers in achieving goals for monitoring and predictive decision-making within the theme parks. This chapter moves back from mandates around wearable technology use and towards choice, as well as how Walt Disney Corporation frames the choice to use wearables as a pleasurable form of interacting with its theme parks.

In 2013 Walt Disney World began implementing a wearable wristband called the Magic Band that promised to allow patrons to move more efficiently through the park space while simultaneously giving Disney purportedly valuable data about the populations flowing through its space. The Magic Band relies on combinations of RFID, other forms of near-field communication (NFC) sensors, and their purportedly frictionless interaction with interfaces and sensors. Ultimately, the discourses about them indicate how they represent a testing ground for the possibility of wearable technologies addressing ongoing problems of infrastructural strain and human population flow in other sites, such as urban spaces. Beneath this cheery imaginary of smiling parents and laughing children, Disney's implementation of the MyMagic+ system entailed significantly overhauling its technical infrastructure as part of an embrace of digital transactions. Basically, wearable technologies are posited as being quite good at performing logistical operations humans and other machines were once asked to do to help make theme parks operate successfully. Mobile media researcher Jason Farman has argued that mobile media broadly help tell stories: geolocative devices like smartphones "give off" information about how and where people come and go, when they inhabit different spaces, how they travel, and what their routines might be.[2] Datafication, in other words, can be interpreted to craft stories about what users do and want, as well as how they behave. Because wearable technologies are presumed to always be attached to a wearer, they help guarantee bodies will be able to interact with beacons, scanners, and stations. These interactions can facilitate access and purchasing power while simultaneously generating massive, real-time reports on population behavior based on datafication.[3]

The building of technical infrastructure is directly connected to the desire to learn about and perhaps even coordinate the behaviors and habits of populations. As I discuss near the end of this chapter, the implementation of the Magic Band offered a means for Walt Disney World to presumably address infrastructural problems in ways that might also be attractive to the developers of so-called smart cities, which aspire to regulate and monitor the daily operations of urban space through automated machine-to-machine communication.[4] Magic

Bands and similar devices intensify existing practices of movement through the promise of a frictionless existence, in which purportedly tiresome human-to-human interactions are rearticulated as supposedly efficient machine-to-machine interactions through the operations of NFCs.[5]

Disney World's Magic Band helps map the park's ongoing and unending state of activities during operational hours, manufacturing real-time data to learn about and respond to trends over time as well as flag anomalies in how groups behave day in and day out. As such, this band helps manage the circulation of bodies and commodities in the park space.[6] The MyMagic+ system demonstrates the transactional nature of data manufacture,[7] as well as the ways in which wearers are asked not to look at the complicated operations of technical systems, instead treating them as gleeful instances of magic facilitating "controlled leisure."[8] Magic Bands are one part of infrastructural systems that strive to learn about populations and adjust daily operations based on data analysis. I first examine what is gained by thinking about theme parks and wearable technologies from an infrastructural perspective, arguing that this provides another vantage point to conceptualize the work institutions do when they engage in datafication. From there, I outline how the Magic Band was promoted, and how it is framed as helping both park patrons and park managers develop ways of managing space. These middle sections emphasize how the Magic Band draws on developments in RFID tagging, frictionless transactions, and other forms of machine-to-machine communication as part of a broader data collection project that is often framed within Disney's rhetoric of magic and wonder in advertising and news features. Finally, this chapter draws on the historical founding of Walt Disney World to position it as an institution invested in experimenting with technologies to help solve problems of urban management. The Magic Band tries to push the datafication of behavior in space as transactional: accessing services or purchasing food and goods creates data to be processed and used in modeling trends of human behaviors. This transactional system is best understood as a form of institutional power, in which data analysis provides suggestions for how spaces (and the people within those spaces) should be managed.

INFRASTRUCTURALISM AND THEME PARKS

The labor involved in producing these datafying infrastructures complicates how theme parks are traditionally conceptualized.[9] Scholarly discussions of theme parks commonly figure them as artificial, highly consumerist spaces. Architecture scholar Miodrag Mitrasinovic, for example, has argued that the success of theme parks depends on their ability to achieve "the totality of the artificial."[10] Even the environment itself is artificial in Walt Disney World, as fake trees and leaves are used as part of decorations.[11] This is to say nothing of the actual environmental harm these theme parks do to their surrounding ecosystems, eroding natural resources and deploying repellants to stave off insects to create an artificial ecosystem.[12] Conceptions of artificiality share affinities with cultural theorist Jean Baudrillard's analysis of Disneyland, which has argued that the space's primary function is to rejuvenate fictions of American ideals by presenting a façade of a sanitary, perfectly functional space.[13] While this line of critique has become quite common, it is also quite limited, failing to account for ways theme parks foreground their own technologies and material infrastructures. If these parks appear as artificial spaces, in other words, it is because of the large amounts of work infrastructural gadgetry performs.

Another way to say this is that the Magic Band reverses this "artificiality critique." No longer are patrons supposed to mistake fake leaves for real; they are instead supposed to revel in the technological efficiencies of the park's infrastructure.[14] While some aspects of that functionality—such as the real-time data collection analyzed within the park's control center to, say, direct the flow of street parades and performers—remain hidden, the act of using the band repeatedly throughout the day constantly announces the presence of Disney World's infrastructure. Not only does this entail substituting human-to-human interactions with computer-to-computer interactions, but it also simultaneously suggests patrons ought to place their trust in computational infrastructure to deliver services and facilitate transactions.

Media studies scholars Nicole Starosielski and Lisa Parks define media infrastructure as "highly automated, relying on sensors and remote control, and requir[ing] human labor for [its] design, installation, maintenance, and operation."[15] Take, for instance, the large and intricate cable systems that help make wireless communication possible.[16] These cables are quite banal; though telephone lines litter highways, mountainsides, and neighborhoods, they are rarely remarked upon (other than perhaps to observe how they get in the way of natural scenery). They have become taken-for-granted elements of networked communication systems. Media infrastructure often emphasizes the capacities for communication between bodies, in either the literal or figurative sense: telegraph lines, undersea cable networks, air traffic control, and satellite systems all facilitate the degree to which words, images, and information all travel at various speeds.[17] As Shannon Mattern usefully noted, the concept of infrastructure "has existed as long as civilization."[18] The etymology of *civilization* provides one way to understand the importance of media infrastructure, and in particular how building and maintaining infrastructure has become a signal of progress and modernization.[19] The French *civilisation*, laden with colonialist overtones, connotes the ordering and cultivation of a population—the opposite, as it were, of barbarity. Infrastructure helps organize, contain, and instill possibilities for the functioning of societies; in other words, infrastructure hurries along the processes of civilization.

Media infrastructure helps build and maintain societies in part because wires, buttons, cables, and stations bind people to one another and ensure that various operations of living—such as electrical currents flowing into and out of houses—occur without (constant) failure.[20] John Durham Peters has suggested that media themselves, not just their infrastructure, are "civilizational ordering devices."[21] For Peters, "media are our infrastructures of being, the habitats and materials through we act and are. . . . Wherever data and the world are managed, we find media."[22] In part through the Magic Band, Walt Disney World is managed through data; the manufacturing of data assists in making a variety of operational decisions each day. Peters's concept of infrastructuralism describes a fascination "for the basic, the boring, the mundane, and all the mischievous work done behind the scenes."[23] The Magic Band participates in infrastructuralism, as the band's operations ask patrons to find wonder in the behind-the-scenes work of infrastructural sensors and automated machine communication. Disney World thrives on an imaginary of "magic"—that which is remarkable, fantastical, and otherwise extraordinary—which hides the unremarked-upon workings of its infrastructure, such as pipes and generators; to borrow from Geoffrey C. Bowker and Susan Leigh Star, "There is a lot of hard labor in effortless ease."[24]

As wearables become more articulated to infrastructural operations like the RFID sensors described in the next section, they become, potentially, more banal and more a part of the everyday operations of moving through these spaces. Like many of the other programs discussed throughout this book, these operations are geared toward datafying the everyday activities of the people moving through these spaces. In doing so, the Magic Band can supposedly assist in making Walt Disney World an even more orderly space that thrives on learning about the habits and behaviors of its patrons.

BUILDING THE MAGIC BAND

The Magic Band was created to increase the functionality of Disney World's various operations. Specifically, the Magic Band focuses on purchases (through linking credit card information to a patron's MyMagic+ account), ride access (through storing and recognizing Fast Passes for different attractions), and crowd sizes (through dispersed sensors that read the location of all Magic Bands in the park). The Magic Band is the wearable component of the MyMagic+ system, which was developed under the project name Next Generation Experience, beginning in 2008.[25] Guests who book a trip through the resort website are immediately offered the option to start engaging the MyMagic+ system to further set up the band. The system asks the patron, among other things, to list their favorite rides, so that their Magic Band arrives prepackaged with an itinerary of suggestions to help them move efficiently through the park on the day of their visit to minimize wait times. If the patron opts in to the band's full range of services through the My Disney Experience website and mobile app, the experience of being in Walt Disney World becomes streamlined through the bracelet: "There's no need to rent a car or waste time at the baggage carousel. You don't need to carry cash, because the Magic Band is linked to your credit card. You don't need to wait in long lines."[26] The promise of frictionless transactions is a sort of alibi for other goals; here, creating faster ways for people to spend money. As much as these bands facilitate interactions with infrastructure, they also help to prompt economic transactions.

The wrist-worn band has the appearance of a Fitbit, although instead of a screen that provides information, the top of the band has a profile of Mickey Mouse. In the park, this representation of Mickey is pressed against corresponding beacons and stations that have the same Mickey profile to activate the band's features. For instance, ride times can be reserved through the FastPass+ feature on Disney's smartphone application or through touching the band to the FastPass kiosk at ride entrances.[27] The band is made of a hypoallergenic plastic and is designed to remain cool against one's skin in the heat and humidity of Orlando, as well as being waterproof to accommodate the various water-themed rides and attractions. It contains a battery that lasts up to two years. While the bands do not have any functionality once a patron leaves the park, they are reusable; the same patron can use the band again when they return to the park at a future date. Any guest who stays in a Walt Disney World Resort Hotel is automatically given a Magic

Band as part of their purchase (the band acts as a room key). It is perhaps better to think about it as wearable infrastructure—a wrist-sized piece of Walt Disney World that patrons carry with them as a constant reminder of the park's services and affordances.[28] There are no buttons, screens, or ways to interact with the Magic Band apart from tapping it against the dedicated beacons in the park. Like many of the technologies discussed throughout this book, it acts more as a worn sensor for manufacturing data or facilitating transactions.

Visitors who do not stay in resort hotels can purchase a band at the park's entrance for a relatively low amount (about US$15) and then sync it to the free My Disney Experience smartphone app. Disney provides free Wi-Fi within the confines of its park to help with this app's functionality. The Experience app sends push notifications and accesses the location of the user's smartphone to keep them aware of FastPass and PhotoPass updates and to allow them to be tracked throughout the park. The app also interacts with restaurants in the park, meaning users can load credit card information onto the app and use it to pay for their meals or order food ahead of entering some restaurants. The benefit for the patron is to expedite the time to receive and pay for food; the benefit for the park is to datafy these transactions and use them as the basis of profiles of patron behaviors and aggregate trends.[29]

These datafication processes also allow patrons to be tagged as they move through the park. This entails NFC through a series of RFID beacons that automatically recognize the unique identifier of each Magic Band as they move past the beacon and in turn transmit that data to the park's control center or relevant employees.[30] For instance, beacons tag each Magic Band as patrons move onto each ride, generating logs of activities. If the ride has a photo taken during it, the photo will be matched to the accounts of bands scanned when the ride began.[31] Similarly, population flow can be monitored to better track lines on different rides and attractions in close to real time. Individual bands and MyMagic+ accounts can be linked together as a "family" to allow for patterns of different families to emerge through comparative datasets, such as evaluating the differences between how families of four and families of six navigate the park, eat, and purchase souvenirs.

Part of the difficulty in implementing the Magic Band system came from the massive scale of Disney's existing infrastructure. Likening Disney World to a "metropolis" more than a park, journalist Austin Carr has noted that Disney World contains "four theme parks, nearly 140 attractions, 300 dining locations, and 36 resort hotels. Its monorail system zips along 15 miles of track, with a daily ridership of more than 150,000. The parks have their own power plant and security force, plus some of the world's largest laundry facilities, cleaning 280,000 pounds of linen each day and dry-cleaning 30,000 cast member garments."[32] Disney has not released information about the cost of this project—to do so would, supposedly, spoil the "magic" of the experience with the knowledge of labor costs—but the immensity of the park's infrastructure suggests that this was both a very expensive endeavor and one that was judged by corporate executives to be of ongoing benefit to the company. Because Disney World uses the same resources as a small city—and was zoned as such when it was first developed in the 1960s—installing infrastructure capable of utilizing the Magic Bands is akin to an urban renewal project.

As one of Disney's "Imagineers"—its own company neologism foregrounding how engineering is actually about producing an imaginary—said during the initial roll-out of MyMagic+ in the early 2010s, "A big part of the company culture is trying to guard against dangerous change. If Disney had followed every trend in the past 60 years, it wouldn't be Disney anymore."[33] Implementing MyMagic+ has entailed designing and constructing new kiosks that can recognize and interact with the bands, rewiring thousands of hotel doors, and creating an entire restaurant, Be Our Guest, that allows patrons to preorder food through MyMagic+'s online applications and then uses the RFID on the bracelets to identify individual guests and alert the restaurant staff about where to deliver food. *Wired* has described how the "magic" of this technology interacts with the park's established infrastructure. Taking the perspective of a tourist family dining at Be Our Guest, the Magic Kingdom's *Beauty and the Beast*–themed restaurant, journalist Cliff Kuang detailed: "The hostess, on her modified iPhone, received a signal when the family was just a few paces away. *Tanner family inbound!* The kitchen also queued up: *Two French onion soups, two roast beef sandwiches!* When they sat down, a radio receiver in the table picked up the signals from their Magic Bands and triangulated their location using another receive[r] in the ceiling. The server—as in waitperson, not computer array—knew what they ordered before they even approached the restaurant and knew where they were sitting."[34] For all this "magic" to be sustained, the Magic Band relies on a combination of short- and long-range sensors throughout the park. The short-range sensors interact directly with the various beacons, stations, and terminals that direct the wearer to touch the band to it, such as cash registers or hotel room doors. Long-range sensors are used more for tracking purposes: "By monitoring where crowds were forming, the company could better optimize flow. Say the sensors noted that one section of Magic Kingdom was becoming overwhelmed with guests: Operators could immediately respond with a character parade around the corner, to disperse traffic and strain on cast members."[35] Short-range sensors, in other words, entail more active participation on the part of the wearer, while long-range sensors are more passive, accumulating information about patrons' behavior throughout their visit.[36]

The Magic Band is emblematic of what surveillance scholars Mark Andrejevic and Mark Burdon have called the "sensor society," which "refers to emerging practices of data collection . . . [to] complicate and reconfigure received categories of privacy, surveillance, and sense-making."[37] This sense making refers to potential reconfiguration of everyday life around various sensors, including wearables. Consider the advertisement for MyMagic+ titled "You'll Want to Use it Everywhere." Here, the Magic Band is imagined as a device that could—and, in Disney's implicit argument, should—become incorporated into everyday interactions with other objects and services. The commercial begins with a man ordering in a coffee shop outside the Disney parks and trying to use his Magic Band to complete his purchase: he absentmindedly presses it against the cash register and the tip jar as an annoyed barista repeats the cost of his order. The commercial cuts to a mother and daughter arriving home with shopping bags; the mother presses her Magic Band against the home's door, expecting it to unlock, only to have both of them collide with the door. From there, the commercial cuts to a mother and son skipping a line at a movie theater, asking for two tickets, and pressing the Magic Band against the glass, much to the confusion of the theater employee. The employee squints her eyes, points at the band, and asks "What is that?"

At this point, the music in the commercial kicks into a more upbeat rhythm and shows the same sets of people performing these actions in the context of Disney's parks: the man buys a stuffed Mickey by touching his Magic Band against a receiver; the mom places her band against a hotel door, opening it; and the mom and son use the Magic Band to gain access to a line for a ride in the park. After the son presses his band against a receiver, a Disney employee says "Hi Chase!" while waving at him; the Magic Band not only proposes ease of use, it also personalizes experience. Chase smiles and exclaims, "She knows my name! This is gonna be awesome!" The narrator confirms: "Once you've experienced MyMagic+ at Walt Disney World, you'll want to use it everywhere."[38] The commercial not only presents Walt Disney World as a more efficient space of quick interactions, but it also positions Magic Bands as having potential to make ordinary operations of everyday life less cumbersome and possibly more personalized. Existing interactions between people and objects—keyholes, credit card readers, ticket-taking patrons—are framed as clunky and inefficient. As the following section details, the "magic" of Disney's MyMagic+ project is the replacement of analog and human relationships with more automated relationships between digital sensors and data collection.

BANDING AND BINDING TO INFRASTRUCTURE

The implementation of Magic Bands at Walt Disney World turns the magical experience on which Disney prides itself into a world of supposedly frictionless interactions to learn about the transactions and movements of patrons throughout space.[39] Each beacon—affixed to cash registers, ride lines, cafeterias, and hotel doors—with which the band interacts creates accumulating moments of capture and datafication. As such, the word *band* is multiply meaningful: it is both a material band, a circular thing designed to sit on a human's wrist, and also summons the Germanic origins of the word, related to "bind." Though now identified as an archaic use of the word, the Magic Band is also "a thing that restrains, binds, or unites," intimately binding the wearer to the infrastructural operations of datafication in Walt Disney World.[40]

While Walt Disney World has always relied on a vast infrastructure to ensure its operational efficiency, it increasingly relies on datafication. For instance, the Magic Kingdom includes an extensive underground infrastructure of cables, pipes, and conduits that helps the park function in ways similar to a city.[41] Similarly, computational forms of infrastructure pulse through the MyMagic+ beacons and devices, akin to what geographers Rob Kitchin and Martin Dodge call "code/space."[42] Computer code, in these spaces, assists with processes for managing the movement of people and objects through space built on making claims about what can be known: what objects have been purchased and how many people have occupied a particular area, for instance. That knowledge about human behavior becomes the basis for making decisions about daily activity, as well as setting baselines and averages over time as populations are analyzed.

Versions of these projects have existed for decades, as Disney has developed and implemented technologies to help patrons navigate the park while developing systems to help learn more about how patrons navigate and use space. In 1999, for example, Disney introduced the FastPass system, which allowed guests to reserve tickets for popular rides, such as Splash Mountain. The FastPass kiosk issued the patron a ticket and instructed them to return to the ride at a particular time. Having returned, the patron could enter an "express" line, bypassing most patrons standing in the queue. The "virtual queue" for the FastPass system is monitored from the Magic Kingdom's "command center," which as of 2012 (right before the implementation of the Magic Band) generated reports every five to ten minutes about the dispersal of tickets and the flow of FastPass lines.[43] The FastPass system helps disperse the population throughout the park by preventing overflow lines. In 2014 the parks introduced FastPass+, which allows visitors to reserve passes to rides up to sixty days before arriving at the park through the My Disney Experience website and app. These FastPasses are loaded onto the patron's Magic Band or, if the patron does not have a Magic Band, a plastic card containing its own RFID tag.[44]

The core idea of the FastPass system—to improve the logistical flow of patrons throughout the park while monitoring activity at different rides and events—informs the design of the My Magic Experience application and the Magic Band. By using this app to prepare a personalized schedule that can maximize the efficiency of their park experience, the My Magic System helps structure the visitor's day to ensure the park's populations ideally move evenly and productively through space.[45] With the Magic Bands, the patron's entire day becomes aggregated as data: their path through the park, the souvenirs they bought, and the food they ate are all available to Disney's Imagineers and their information processors in ostensibly real time as part of the larger aggregation and accumulation of information about how the Disney parks function. These analytics were originally developed to create "forecasts" of park operations. In 2012, for example, the Magic Kingdom generated "transaction forecasts" for every fifteen-minute period of the day, gathering data through park entry turnstiles, restaurants, and gift shops, with an eye toward ensuring that the park's labor resources were suitably dispersed.[46]

Games scholar Ian Bogost, reflecting on his experiences with the Magic Band, suggested that it "lays bare the process by which we produce data—not all on our lonesome, but as the result of implicit and explicit pacts with organizations, most often corporations."[47] Datafication at Disney World, in other words, exists to help corporations optimize services, in much the same way recommendation systems like Netflix monitor metrics pertaining to not only what but *how* viewers watch content.[48] Just as Netflix's engineers treat users as test subjects in a wide range of experiments to refine the company's recommendation algorithms, so too can Magic Band be implicated as an experimental device meant to refine and adjust spatial operations through a blend of machine and human learning.[49] The "magic" around Magic Band is also meant, then, as a means to lure patrons into doing the labor of manufacturing data for Disney's engineers and managers to evaluate.[50] In performing such optimizations, in other words, Disney relies on claims about how datafication can lead to understandings of human behavior at population levels.

THE FRICTIONLESS KINGDOM

The Magic Band facilitates infrastructural operations in part through the concept of frictionless sharing, which supposes a seamless sharing of information between technological devices that facilitates transactions and interactions across space. When a wearer touches their band to a FastPass beacon and the beacon acknowledges their FastPass and grants them access to the ride queue, a frictionless transaction has occurred. Two machines—the band and the beacon—interact to determine the user can access the ride, but if this infrastructure works according to plan, these interactions barely register for the user. The frictionless sharing of information this band facilitates has been made somewhat popular through applications such as Apple Pay. Apple Pay, like the Magic Band, removes the presumed inconvenience of materials like credit cards by allowing users to store credit card information on their devices and, rather than inserting a card into a reader, open the Apple Pay application and hover their device (iPhone or Apple Watch) over a designated pad to complete the transaction.

Friction, in other words, happens when people encounter other people or devices—magnetic strips going into card readers, cash going into the hands of clerks, or pens writing on the surface of checks. The word comes from the Latin *fricare*, to rub; to remove rubbing implies a unified, smooth system. Removing friction at the point of purchase makes economic transactions happen "instantly," without the purportedly complicated resistance of human interaction.[51] To be frictionless implies efficiency and speed, coordinated through technical systems.[52] The term operates as a stand-in for processes of machine-to-machine communication, masking the infrastructural operations that allow otherwise "invisible" communications between devices to facilitate transactions. When they are successful, these infrastructural operations smooth out processes of transaction, access, and other aspects of moving through space.

Friction, as a concept, calls attention to the minute processes comprising infrastructural operations, as well as how humans use technologies to perform tasks such as purchasing food and souvenirs. Friction, from this point of view, is not a fact but rather a value, and one that shifts historically; that which is judged to be cumbersome depends in some part on the wider technological assemblages facilitating any given action or practice. As columnist David Pogue summarized, "Friction is a hassle. Steps. Process. And in this increasingly technified world, there is still a surprising amount of red tape—and few examples of push back. We stress about things like price, storage and processor speed, instead of beauty, elegance and low friction."[53] Pogue's reference to process is important, suggesting that as technologies become institutionalized, the relationship between friction/less is reimagined. Such celebrations of a frictionless world overlook how engineering the removal of friction has entailed, at least in the 2010s, the transference of protocols from people to machines through considerable human labor, maintenance, and governance.

Consider, as an ancillary example, so-called frictionless spending. When Apple announced Apple Pay, a feature allowing users to pay via interactions between an iPhone or Apple Watch and an Apple Pay–equipped beacon, it produced a video focusing on a woman making a purchase at a cash register. The video isolates each moment of interaction between her and the materials surrounding her transaction: her purse, her wallet, her credit card, her identification, the credit card scanner, and the salesperson facilitating the transaction.[54] Apple's video emphasizes speed, suggesting that the removal of these objects—and their replacement by an Apple iPhone or Watch and a sales beacon—fundamentally alters the point of sale. And to Apple's credit, points of sale have indeed changed over time: cash register checkouts allowed for an expedited sale because they provided ways around written invoices and hand calculations; self-checkouts get around interactions with a cashier; debit cards get around writing a check.[55] Interactions are not really eliminated here; they are replaced. Rather than two humans exchanging material money and tracking prices, two (or more) technological devices exchange information to facilitate the transaction in ways that only appear to be invisible.[56] Similarly, the My Magic Experience system works to load as many interactions as possible onto the application and the wearable.

In a bit of a paradox, the proposed removal of friction also depends on the appearance of ever more technological apparatuses, both visible and behind the scenes. Google incorporated a clip from the television series *Seinfeld* to initially advertise its attempt at a similar application, Google Wallet, in 2011. In the clip, George Costanza (Jason Alexander) stumbles across a sign that informs him he can use Google Wallet to turn his phone into his wallet. He pulls out his wallet, which is comically overstuffed with receipts and scraps of paper.[57] The joke here is that the analog world is full of clutter, but the digital world simplifies, streamlines, and cleanses.[58] Despite the promises of such advertisements, frictionless sharing is more an ideal than an actuality. Magic Bands still interact with portals, bases, and stations that need to read identifying information and respond in kind to appropriate commands. To mobilize "frictionless" engagements is to turn attention from the communication infrastructure that operates in the background. When the infrastructure fails—when a card reader fails to register a credit card, or a cell tower goes down and lowers the quality of cell-phone service—friction is reintroduced.[59] The capacities for frictionless activities relate to the capacities of infrastructure to sustain such activities.

Devices like Magic Bands, in other words, rely on an imaginary of "frictionless" human-machine relations. They require some sort of fastening and grasping against the wearer's body, as well as the ability of the wearer to move to and interact with the contact bases. These bands replace (or in some cases, complement) interactions with park employees with the Mickey-shaped beacons lining the park, substituting human interaction with sensor transaction. Interactions do not disappear; they are transformed. To call them frictionless is to presume the removal of process; in reality, the processes that normally govern human-to-human interactions have been relegated to the often-passive ways the devices speak to each other through radio frequencies, NFC, and other types of sensors that trigger the exchange of information and datafication of the wearer's activities.

Alongside the concept of friction, the Magic Band can be understood through the related concept of *seamlessness*, which has been one of the project's major buzzwords. Disney's chief operating officer, Tom Staggs, said in 2013 that the band would create a more "seamless" and "personal" park experience, and that language was later employed by travel writers and technology bloggers alike to describe using the band.[60] The differing definitions of the word *seam* are helpful to consider in this regard: the word can mean "the joining of two pieces by sewing," "a thin layer of stratum between distinctive layers," or a "vulnerable area or gap" depending on the context.[61] Infrastructure depends on labor and maintenance to remain operational.[62] In Walt Disney World, the Magic Band joins together multiple different forms of interaction while simultaneously producing data meant to maintain infrastructural operations and prevent breakdowns.[63] Implementing these technologies works to make machine-to-machine communication a more habitual practice that accompanies moving through public space. As the following section makes clear, implementing this sort of technical system is in line with how Disney's Orlando parks were originally conceptualized as spaces to experiment with urban design, practice, and operations.

INFRASTRUCTURE AS DIAGRAM

When Walt Disney formally presented the concept of the Epcot (Experimental Prototype City of Tomorrow) Center to the Florida legislature via a posthumous video in 1967, he remarked that there was no greater challenge anywhere than finding solutions to the problems of cities.[64] Over fifty years later, the Walt Disney Corporation continues to position its parks as places to implement technology that might solve problems endemic to urban settings.[65] In the 2010s, one of those problems was framed as monitoring population flow through, within, and across spaces, and related issues of administrating population management and security.

The Magic Band's relationship to friction and datafication is, in this context, part of a much larger diagrammatic project, one that models the relationships between technological objects, bodies, and space for human and infrastructural management. Here, I mean to reference philosopher Gilles Deleuze's notion of "diagrammatic," which entails relationships between, among other things, the form and function of objects. In his writing on Michel Foucault, for example, Deleuze described a how a "thing"—like, for Foucault, a prison—becomes both an existing "formation" (a thing in the world) and a form for expressing ideas.[66] To invoke one famous example from Foucault's writing, prisons are not just formed in space; they express ideas about how to discipline populations. Somewhat similarly, theme parks like Disney World are not only formed in space; they express ways to imagine the articulation between technology and infrastructure. Spaces like Disney World can model ideas about potentially emergent ways of governing everyday practice.[67] They offer diagrams for other companies and urban developers to potentially draw upon for the construction of similar infrastructure.

Since its founding, Disney World has positioned itself as a testing space that can help solve the problems of cities through the progressive development of technology. In late 1967 Disney (the corporation) presented its development plans to the Florida legislature, using a pre-made film posthumously featuring Walt to sell the idea of Epcot. In the video, Walt described Epcot as a model city that would "take its cue from the new ideas and new technologies that are now emerging from the creative centers of American industry. It will be a community of tomorrow that will never be completed, but will always be introducing and testing and demonstrating new materials and systems. And Epcot will always be a showcase to the world for the ingenuity and imagination of American free enterprise."[68] There is much that suggests the emergence of a hyperrational control society in Disney's speech, drawing on futuristic promises to create an always-already unfinished space that is autonomously controlled by the Disney corporation.

Beyond the confines of the theme park, Disney's interest in exercising control over space is evident in Celebration, Florida, a town the company began developing in 1991 to, in part, deliver on the promise to create a fully inhabitable community at Epcot Center. Celebration's outward façade is governed by a strict set of regulations that make the town appear to be "trapped" in the early twentieth century. The town's official website calls these "Design Guidelines," and provides "templates" for how to properly create landscapes, awnings, lighting fixtures, fences, and other exterior elements to conform to the town's overall aesthetic.[69] Celebration attempts to recreate a nostalgic small-town America, a wholesomely knit community that thrives on neighborliness. As cultural studies scholar Andrew Ross demonstrated in his ethnography of Celebration, the everyday life of the town has a difficult time living up to the planning: things break down, problems occur, and other unplanned incidents get in the way of the control the town seeks to administer through things like design guidelines.[70] The guidelines demonstrate Disney's ongoing commitment to seamlessness, but everyday life is, indeed, a seam-ridden domain, full of unexpected ruptures, evasions, and departures that break from the efforts to establish control.

Epcot's design and function are not dissimilar from Celebration's, in that the engineers and managers who run the park are also engaged in an ongoing process of trying to figure out ways to control the uncontrollable messiness of living through technology. While Epcot was founded on showcasing, testing, and implementing emergent technologies, urbanist Steve Mannheim's suggestion that its design tried to resolve the "diminished importance of downtowns in many American cities" underscores the relationship between the park and other public and semipublic spaces.[71] Disney "was quite serious initially about the word *experimental*" in Epcot's anagram and believed Epcot would offer testing grounds for designing ways to live.[72] The experiments of Epcot—and Walt Disney World as a whole—are not just about the showcases on the surface and the infrastructure pulsing below the sidewalks, but also about giving patrons the opportunities to have experiences with emergent technologies. Just like the My Magic Experience+ commercial suggesting that patrons will want to use their Magic Bands everywhere, magic is accomplished when what begins as a new sensation of technological fascination and wonder transforms into the expected and banal.[73]

To further position how the Magic Band advances the larger project and history of Epcot and Walt Disney World, it is helpful to think about how the design of the park fits into theories of space. Early sketches of Epcot show Disney relying on a "radial plan," wherein activities and "neighborhoods" would be organized around a central hub—borrowing directly from the design of the Magic Kingdom, which is organized around the main node of Cinderella's Castle. This spatial configuration "dates at least to the first century B.C. The problem with such plans over time is the difficulty of keeping growth within the boundaries."[74] Disney's—first Walt's and then, after his death, the corporation's—plans for Disney World also argued that the parks be "freed from the impediments to change, such as rigid building codes, traditional property rights, and elected political officials," even as they were trying to position Disney World as an autonomous political district for purposes of districting, zoning, and building.[75]

This is a problem Michel Foucault explored in his work on territories. For Foucault, focusing mostly on historical changes in Western Europe, the enclosed nature of towns began to change in the eighteenth century, expanding and creating questions about "the spatial, juridical, administrative, and economic opening up of the town: resituating the town in a space of circulation."[76] While spaces like Epcot lack the same juridical and economic roles as other towns—in that it is a corporate-owned space comprising visitors and employees, but not residents—the Magic Band's capacity to monitor how things circulate through this space demonstrates how there is an overlap of concerns between how the park is managed and how urban spaces are managed.

This blur between the challenges of managing a leisure space like Walt Disney World and the challenges of managing urban spaces more generally appears elsewhere in theories related to space. Social theorist Henri Lefebvre argued that the latter half of the twentieth century was invested in creating "architectures of enjoyment," or nonwork spaces designed to exclusively facilitate pleasure.[77] While Lefebvre's work on this phenomenon—which he called *jouissance* as a sort of umbrella term—focused more on the proliferation of European tourism, rest, and leisure, it is not difficult to draw a line to Walt Disney World, especially because Disney sought and successfully received special districting for its Florida properties. Disney World was always conceived of as a series of cities rather than a series of parks; they were built both for *jouissance* and to serve as experimental civic models. In his celebration of such spaces, Lefebvre proclaimed: "Listen to the sound of your footsteps, the muffled timbre of your voices that accentuates the slightest intonation, the intensity of birdsong, the grass, the rain within the space of a cloister. There is something to be learned here."[78] To Lefebvre's poetic descriptions of the promenade, one might now add: try to hear the frequencies of radios, the invisible passing of data, the click of beacons registering each passerby.

The Magic Band—and the rewiring of Walt Disney World that accompanied it—serves as the latest event in a decades-long project of using the park as a testing ground for managing populations in a contained environment. In the 2010s, computer information technologies promised to datafy park transactions in real time, offering a strategy for monitoring and managing these messy and unpredictable flows. The infrastructure of fiber optic cables installed throughout these spaces, the ability of satellites to engage GPS in real time, the implementation of NFCs, and the proliferation of RFID work together to generate aggregate analytics of populations and materials flowing in, out, and through spaces. These technologies promote a transactional view of space, in which stations and turnstiles are accessed and experienced through swiping or pressing key cards and bands against or close to surfaces.

The Magic Band's administration of a transactional space signals the emergence of a technical diagram binding institutions and individuals together through the infrastructural work of bands, sensors, chips, and scanners. Public space's features "are now subject to computational enhancement. Street furniture such as lamp posts, signage, even manhole covers can provide the urban sojourner with smart waypoints."[79] Management information professor Anthony Townsend defined so-called smart cities as "places where information technology is combined with infrastructure, architecture, everyday objects, and even our bodies to address social, economic, and environmental problems."[80] Smart cities depend on relays of information between individual devices and infrastructure designed to both "read" these devices and provide individuals continued opportunities for transactions to access, purchase, and consume services and spaces.[81] Smart cities like New Songdo, South Korea, use millions of sensors in their "roads, electrical grids, [and] water and waste systems to precisely track, respond to, and even predict the flow of people and material."[82] Rather than relaying information about bodies to cultivate social capital, as do the fitness trackers discussed in chapter 1, devices like the Magic Band connect bodies to built environments and their administrators.

While spaces like New Songdo are more the exception than the rule at the time of this writing, they also offer, much like Walt Disney World, an aspirational sense of what the smart city could look like and how it could function. Researchers like Aaron Shapiro have argued that beneath the magic of technological seamlessness is a system of ongoing extraction.[83] Companies like Uber and Lyft—which track every second of their employees' shifts—may not immediately read as components of smart cities, but for Shapiro they are inextricably part of the push to datafy and monitor the real-time movement of people and goods through urban space. Local institutions may introduce RFID tags, scanners, and readers "to improve residents' 'quality of life,' provide increased convenience, and ensure better maintenance and efficiency of city and facility service,"[84] but this convenience comes at the cost of constant datafication. The data manufactured to facilitate this convenience is also often one-sided and favors those who have some form of power over space, be they urban engineers, building owners, heads of public transit, or restaurateurs.

As has been argued in assessments of how Disney collects data on users and how its privacy policies frame data collection, privacy is becoming increasingly contextual, and its norms of governances shift as users are asked to consent to near-constant monitoring in various spaces.[85] One legal argument has suggested that "the information that is being collected and used by the company is the type of information that could allow a person to create a detailed and descriptive profile of the individual guest that amounts to more information than what the individual is able to express by simply being in public."[86] Via the Magic Band, as with the many proliferating components of smart cities, what it means to exist in public is increasingly articulated to surveillance and monitoring technologies. As opposed to things like closed circuit television cameras and other forms of public surveillance, which have been in place for some time, the communication about a person's behaviors and actions via machine-to-machine communication and computational infrastructure creates new problems for how to define being in public as an act of contributing to datafication. Walt Disney World may tout its "magic" as the alibi through which purportedly new and exciting experiences are made possible, but this magic's alchemy has taken a toll on how space might come to be understood as one more component of how the ideology of datafication presumes that monitoring and manufacturing all these various transactional data is a pathways to better understanding of human behaviors.

CONCLUSION: THE INFRASTRUCTURE'S MAGIC TRICK

This chapter has explored how wearable technologies are articulated to infrastructures to assist with population management. Magic Bands and their sensor beacons offer aggregation-based strategies for learning how populations use space. The creation of real-time data about population flow promises to solve how to process and regulate circulation through space. This infrastructural system attempts to institute new modes of ordering built on discourses about seamless and frictionless transactions, rendering space both transactional and constantly extractive. As this chapter has explored, sensor-based infrastructure records transactions and movements, allowing companies like Disney to build granular understandings of human activities and use those to guide interpretations of human behavior. Media and communication scholar Alexander Galloway has contended that "in order to be in a relation with the world informatically, one must erase the world."[87] As this chapter has argued, the concept of frictionless technology supposes an erasure of interaction, when in reality it substitutes some forms of interaction with others, transforming spaces into sensor-governed systems.

The world, of course, is not being erased so much as rewired. Disney calls this rearticulation of interaction "magical," and that certainly fits with one of the traditional genres of magic trick: making something disappear. Like a good magic trick, accepting that something *has* disappeared—like a rabbit into a hat—requires a bit of suspension of disbelief. It requires one to *refuse to look at* the infrastructure of the magic trick; the engineering and labor that went into the magician's craft make the trick appear to be seamless, even though there are many stitches and seams that must be perfectly coordinated for the trick—like infrastructure—to not break down in its process. Infrastructural components are magic tricks in that they only *appear* to erase the labor that sustains the world. As a magic trick engineer remarks in the 2006 movie *The Prestige*, "The magician takes the ordinary something and makes it do something extraordinary. Now you're looking for the secret, . . . but you won't find it, because of course you're not really looking. You don't really want to know. You want to be fooled." With the Magic Band and similar kinds of technologies that increasingly replace keys and cash and money, the greatest trick they pull is that they too come to be seen as ordinary things that make the world meaningful.

Conclusion

CULTURE, OR: ORDER AS A WAY OF LIFE

THE SIX CHAPTERS comprising this book have mapped different articulations between wearables and a broad array of sites, representations, and debates. My analysis has focused on discourses *around* wearables, using public documents to illustrate some of the processes, policies, and programs that have connected wearables to law enforcement, infrastructure, health monitoring, sports, labor, and accessibility. I have also attended to the tensions between choice and mandate in wearable technology use and how the forms of power wearables help express complicated relationships between wearers and different systems and institutions. In the conclusion, I want to discuss more directly how I have used discourses around wearables to tell a larger contextual story about changing existential conditions and political possibilities in the face of everyday life's datafication. As I have argued throughout this book's various examples, wearable technologies operate alongside and through policies, protocols, ideologies, and governance to manufacture not simply data but potentially new ways of life.

Cultural studies scholars Jennifer D. Slack and J. Macgregor Wise proposed the term *technological culture* to describe the linkages and connections between technology and culture. For them, technological culture describes deep connective threads that mutually constitute culture and technology as discrete existential realms, indicating how each has been articulated to the other so thoroughly that disentangling them is nearly impossible.[1] Another way to say this is that technologies participate in and increasingly cut across fabrics of culture, society, and economy; they are not isolatable devices, but always-already stitching together discourses, practices, and policies shaping their imaginaries.[2] This conclusion, then, recasts the book through returning to some of the theoretical and conceptual language used in the introduction, as well as providing some possible paths for challenging forms of the datafication of everyday life I have critiqued throughout these chapters.

THE UNITY OF DATAFICATION

This book has primarily used the discourses around wearables technologies to consider a series of articulations and maps of connections between the material objects of wearable technologies, as well as the representations, practices, policies, and institutional uses of those technologies that come to bind together the domains of technology and, say, health. As Lawrence Grossberg once put it: "Articulation is the production of identity on top of differences, of unities out of fragments, of structures across practices."[3] In the ways wearables promote and produce normalcy through things like step measurement, in the ways employers or supervisors use them to surveil and compel individuals into expected behavior, and in the ways they are imagined to solve extant social problems, the connective threads of this book indicate the production of a "unity" around the supposed value of data manufacturing. The datafication of everyday life is the name of this unity.

The map of articulation is meant to glimpse what cultural studies calls an *assemblage*, which comes from the philosophy of Gilles Deleuze and Félix Guattari and speaks to the dynamic and ever-shifting form of accumulating articulations.[4] Articulations and assemblages are never settled things. The lines on the map are constantly moving, and tendential forces become stronger and weaker over time. Similarly, an assemblage may take different forms over time. As Slack and Wise defined it, "an assemblage is a particular constellation of articulations that selects, draws together, stakes out and envelops a territory that exhibits some tenacity and effectivity."[5] Without getting too into the weeds of Deleuze and Guattari, suffice it to say the vocabulary they provide describes how assemblages change form (deterritorialization) and how new connections or articulations are forged within existing or emergent assemblages (reterritorialization).[6] The datafication of everyday life—the unity produced through the connective fragments of this map—is something close to an assemblage. It is a formal configuration of articulations between data-manufacturing technologies and existential social territories and institutions, and it has evident political consequences in its operations. When I speak of politics here, I mean (among other things) the distribution of resources, the shaping of public life, techniques of governance, means of existential support, and sites of power and contestation.[7]

These articulations—these lines of tendential force—can become something resembling hegemony if they become settled enough. As Antonio Gramsci developed it, *hegemony* is a theory of power that describes how individuals are governed through consent rather than coercion, or how the production of common sense relies on internalizing some element of power.[8] Gramsci's key point—a point he himself quite literally embodied as a political prisoner in Mussolini's Italy—was that hegemony could never be a thing achieved but always something in process of becoming. Individuals always have room to resist power, to circumvent it, or to build their own pathway creatively and tactically through it.[9] This concept is crucial to Michel de Certeau's practice of everyday life, in which individuals use tactics in response to structures meant to govern how they can live.[10] In de Certeau's work, tactics are inherently limited things; his *Practice of Everyday Life* is built around a theory of "making do," of trying to understand how people make sense of a world where they may have little agency to change structures of oppression, yet they continue to try and live well all the same.

One of the political and analytical goals of cultural studies—beyond understanding the complexity of "what's going on" at a particular conjuncture in space and time—is to figure out how to produce change, to "consider where there might be lines, connections, relationships, and articulations that could be altered, where the lines of force are less powerful, more vulnerable."[11] Nearly every case study in this book entails some entity or institution trying to render the datafication of everyday life as a hegemonic force—that is, assumed, accepted, and strategically dominating the terrain of the everyday. But as Ben Waber was quoted as saying in chapter 4, culture seems to keep getting in the way. De Certeau's conceptualization of the rigidity of structures also helps frame how difficult it might be to transform the intensifying datafication of everyday life. Nevertheless, we collectively must look for creative and tactical paths to cut across the seductive or coercive lure of datafication's ideologies and practices.

DATAFICATION AS ADMINISTRATIVE CULTURE

To turn away from the admittedly dizzying theories of articulation and assemblage, let me connect this conversation more fully to *culture* as an existential domain. Cultural historian Raymond Williams famously described culture as, in part, a "whole way of life."[12] For Williams, culture entails an interplay between artifacts (for this book's purposes, wearable technologies, but also an array of material items that comprise everyday practice), values (datafication and other ideologies sympathetic to data manufacturing and analysis, rationalization, and administration), and collective forms of living based on a negotiation of established and emergent practices.[13] For Williams, culture was always in flux, between known and unknown directions, between accepted patterns of conduct and experiments in living. This book has charted how wearables are part of an ongoing transformation of culture characterized by changing forms of administration these technologies have assisted in producing.

This tension between what we might call a top-down, prescriptive culture and a bottom-up, ordinary culture has been festering for centuries. As Theodor Adorno once put it: "Whoever speaks of culture speaks of administration, as well."[14] Culture's etymological and social ties to civilization has tethered it to imperialist and elitist projects of maintaining the status quo, and projects of cultivation have perceived culture as the arena of disciplining various bodies (youth, immigrants, "deviants") into acceptable modes of comportment.[15] Ted Striphas, in a recent history of algorithmic culture, has argued that not only are "culture and computation" deeply entangled modes of conceptualizing rationality, they also have the potential to "underwrite oppression."[16]

Technologies contribute to tensions between how individuals might organically use different tools to shape their conduct and how various vectors of force prescribe and insist on technologies as necessary for administration to function. Indeed, Lewis Mumford is among the many scholars who perceptively saw the relationship between "civilization" and technologies, in which each mutually shapes the other in the development of social affairs.[17] In other important formulations of what could be called technological culture, uses of technology reflect broader political, social, and moral imperatives of human cultures at different points in time, from Friedrich Kittler's analysis of "discourse networks" to Harold Innis's understanding of communication infrastructures shaping cultural relationships to space and time.[18]

Analysis of technological culture has not just considered how technologies (as material artifacts and systems) and culture (as fabrics of meaningful social reality) are co-constitutive. It has also considered shifting modes of governance, mutations of power that continue to produce forms, ideas, and arrangements of human affairs. In 1967 Henri Lefebvre coined the phrase "bureaucratic society of controlled consumption" as the core contribution of his book *Everyday Life in the Modern World*. Lefebvre lamented that a sense of everyday life as a "non-philosophical" domain has been replaced with growing regimes of rationalization, programming, alienation, and reorganization, such that everyday life becomes envisioned as a "structured and closed" system.[19] As Lefebvre said, "Everything here is calculated because everything is numbered: money, minutes, metres, kilgrammes, calories . . . ; and not only objects but also living, thinking creatures, for there exists a demography of animals and of people as well as of things. Yet people are born, live and die. They live well or ill; but they live in everyday life . . . it is in everyday life that they rejoice and suffer; here and now."[20] Lefebvre's complex portrait of everyday life at a crossroads between different systems of imagination and accounting underscores how technological cultures are also and always administrative cultures, and their purview cuts deep into the heart of how people imagine their conditions of possibility for managing and navigating everyday life.

BRINGERS OF CHAOS

The datafication of everyday life is one component of what I called in the introduction a computational Age of Order that may be defining technological developments of the last quarter to half century. Order is a vector, a force exerting itself on how institutions and individuals imagine daily life. Let me return to the quote that opened this book: "The goal of wearables . . . is to help bring some order to life." I have used this quote as my own sort of ordering mechanism, arguing that it expresses the logic of how articulations between wearables and different domains have been built and reinforced. However, the quote also contains an implicit judgment that *order* is inherently valuable. While key theories of everyday life like habit and domesticity have always been about the cultivation of organization, these were more organic care structures developed by individuals to make their way through the world, rather than administrative modes of governance to monitor the efficacy and health of various habits.[21] What the proposed goal of bringing order to life suggests is that mundanity *is* chaos, and it needs to be organized. Or, to put it in language used previously, subsumed into the assemblage of datafication.

Deleuze and Guattari suggested everyday life is replete with "antichaos forces," efforts at minimizing the infiltration of chaos through "selection, elimination, and extraction."[22] As each chapter has explored, different institutions and companies try to reduce chaos: for the health monitors discussed in chapter 1, the COVID-19 pandemic created chaotic conditions in which manufacturers could position their devices as helping through an orderly diagnosis of bodily conditions. For the body-worn cameras discussed in chapter 5, the violence of police brutality created conditions in which cameras were imagined to rearticulate "law and order." In other stories told in this book, the relationship between order and chaos has more economic ends: for Humanyze and Walt Disney World alike, wearable sensors offer a means of ordering workers, patrons, and populations. The hearables in chapter 2 held a promise of organizing sound data for personalizing individuals' experiences. Each of these articulations can be interpreted as the suppression of chaos and the instantiation of some form of order through the datafication of everyday life. Wearables are part of the history of media technologies devoted to recording human life, acting as knowledge machines.

I have walked (albeit briskly) through this terrain of critical and social theory to arrive at a relatively straightforward point: we must not brush past the importance of recognizing that daily life is quite chaotic. While scholars of everyday life have long noted the importance of crafting habits, building domestic spaces, structuring routines, and balancing repetitious actions, these all exist in a dance between order and chaos, a dialectical production that has structured this book's arguments about the discourses surrounding wearable technologies and their role in broader administrative projects.[23] The datafication of everyday life is one potential product of this dialectic, a synthesis that risks defining "the chaotic" as the detritus of everyday life. While I do not reject whole cloth the building of order into my daily life, I want that order to exist on my own terms, rather than be prescribed to me. I want protection over the data of my everyday life. I want to decide to implement a body-worn tracking device. I do not want to be the object of attention of a surveillance mechanism. I want to move through public space without my footsteps becoming calculable to others. Perhaps some of these requests sound naïve, but then it is worth asking *why* they sound naïve. If we are not granted the opportunity—nay, the right—to refuse the datafication of everyday life, then we risk continuing to cede the ground of our own daily lives to governments, corporations, and other institutions that embrace the territorializing capacities of datafication. If we cannot turn the tide of datafication, we must at least build ships capable of navigating these waters.

This book's mode of critical inquiry has been persistently and insistently oppositional to the current strategies at work in building articulations between wearable technology and social domains. It has pointed repeatedly to the ways they serve to surveil, marginalize, and enforce regimes of power—sometimes even as they claim to solve problems and make life better. But the project of critique is *not* one of mere opposition; it is one of transformation and change.[24] Here are some ways we might try to treat wearable technologies as chaotic technologies and begin rearticulating what we can do with these devices:

1. Reject technology prescriptions. Chapters 3 and 4 (sports and labor) focused on the problems of using wearables to surveil student-athletes and workers, respectively. Along with the use of wearables in life insurance plans discussed in chapter 1, we might broadly call these technology prescriptions, moments at which someone in authority over an individual compels them to wear a device for their own good. We should develop more careful policy building and analysis here and examine whether these prescriptions are merely excuses to enhance surveillance and apply pressure to individuals. Where possible, individuals should speak out against these policies in public, demand their data be deleted, or choose (if possible) to not engage with these systems.
2. Subvert advertising as guides to how *not to* use new technology. In chapter 2 (accessibility), I discussed how Apple's advertisements for the AirPods focused on the affordances of mobility and musical playback rather than the affordance of accessibility, thus reinscribing into the national advertising campaigns normative understandings of whom these technologies were for and what they could do. Advertising for new technologies often acts as a space of constructive representation, showcasing potential affordances and routes of use. One way to treat technologies chaotically is to refuse to use them as advertised and instead engage in playful exploration of the features and settings to discover how they have been preprogrammed and what other pathways of use may exist.
3. Understand technologies are techniques for living. Here I am combining the need to understand technologies as *techne*, as related to practical skill and ways of doing things. As I discussed in the introduction and a bit in the preceding section, wearable technologies are geared toward everyday life and building habits, documenting routines, and reflecting on what our bodies do and how they do them. In the years I have been writing about wearables, I have had many people tell me they find them

helpful and useful, and there is good reason to take their comments seriously. But rather than encourage systems and institutions to deploy and prescribe them, we should reinvest focus in seeing wearables as means to help people live their lives and live well. Student-athletes, for example, should have the right to monitor their own sleep and learn from whatever that data might tell them without automatically ceding it to their coaches and trainers for review.

4. Improve collective literacies around what, exactly, algorithms, storage systems, and data analytics are and how they work. While this book has not been a technical manual, it has routinely pointed to moments when technologies have been dramatically simplified in public communication. For instance, the claims that Oura Rings could detect COVID-19 symptoms were vastly overinflated, and the "magic" of the Walt Disney World Magic Band does not explain to park visitors how near-field communication or radio frequency identification works in any useful way. The hope that police body cameras would improve transparency rested on an idealized understanding of how evidence storage interacts with the law. I am under no illusion that literacy in and of itself can solve this problem, but it provides one strategy for helping people understand what these technologies are and how they work. As suggestion 1 asks people to speak out against policies in the hopes of applying pressure on institutions, so too can providing more public literacy create the conditions for people to hopefully speak about these concerns where they matter.

5. Be ambivalent toward our own data. Part of the power of datafication as an assemblage and an ideology comes from the seductive lure of the empirical, the possibility of accessing some inaccessible realm of knowledge through converting it to "data." While collecting and analyzing data can be important and useful for building knowledge, we should always be treating the data manufactured through wearables with a degree of ambivalence, as containing both affordances and limitations.[25] I have worn different kinds of wearable technology in the decade I have been doing this research, from Fitbits to GoPros to Apple Watches to blood oxygen rings. I treat them with skepticism, but I cannot deny they have produced moments of interesting reflection and consideration on my own condition. Generating more collective ambivalence toward datafication could help dislodge some of the ideological fervor around it.

6. Join collective forms of action in established groups. A variety of organizations have been formed to develop agential collective responses to the injustices that can be exacerbated through technologies. These include the Algorithmic Justice League, the Global Indigenous Data Alliance, and the Distributed AI Research Institute.[26] Becoming engaged in this work can provide avenues for advocating for reform or transformation in the implementation of technologies and datafication.

These suggestions cannot, in and of themselves, solve the problems outlined throughout this book, but they may begin processes of disarticulation that can put critique into action and begin transforming the ways we talk about, imagine, and ultimately argue for or against the datafication of everyday life. Lefebvre, for his part, suggested a "cultural revolution" aimed squarely at undoing the bureaucratization of everyday life. Writing in the late 1960s, on the proverbial eve of the May 1968 revolutions in Paris, revolution was in the air. Several years later, in 1972, Michel Foucault and Gilles Deleuze reflected on the role of "the intellectual" in the aftermath of the 1960s' efforts at social and political change. Foucault put it this way: the intellectual struggles "against the forms of power that transform [them] into its object and instrument in the sphere of 'knowledge,' 'truth,' 'consciousness,' and 'discourse.'"[27] Foucault would add in a different interview: "There is a battle 'for truth,' or at least 'around truth,'" where truth is understood as a "thing of this world," made through the world, rather than some universal myth of "truth" that can be accessed or attained.[28] The sphere of "knowledge" and "truth" has transformed considerably since 1972, but Foucault's astute observation that there are battles over the nature of truth continues to reverberate.[29] It is a crucial part of this book's conceptualization of wearables as knowledge machines and of datafication as an assemblage built on computational analysis of human conduct. The tactics I offer here are responsive and reactionary, and they will likely look different for each of us, but they provide a path for reconsidering how datafication exerts a rearticulation and redefinition of everyday life.

MOMENTS OF CHOICE

As I mentioned at the start of this conclusion, articulations are always ongoing and always multiplying. I am finishing this in March 2024, and by the time this book is published and reaches your hands, things may look considerably different. Technological culture moves quickly, and nothing is assured. In sketching articulations between wearables and different domains, I have focused mostly on consumer-grade wearables (like the Fitbit or Apple Watch) and categories of devices that have generated considerable public discourse (like body-worn cameras or workplace surveillance badges). There are more histories to write of the highly specialized wearables developed for health-care settings,[30] other uses of wearable cameras in sports settings,[31] the development and delayed promise of virtual reality headsets and augmented reality glasses,[32] and surely dozens of other categories of devices. They speak to other articulations, other threads and fragments whose connection to the domains considered in this book may illuminate further the strategies of tethering wearable technologies to modes of power.

Raymond Williams once said, "Every new technology is a moment of choice."[33] This book has examined a series of choices made around the emergence of wearable technologies, many of which have inscribed them into existing regimes of power and articulated them toward the ongoing datafication of everyday life. The map of articulations this book has traced has examined choices in discourse, representation, policy, and administration that have all contributed to transformations over how different people are asked to conceptualize their habits and their everyday lives. While critique as a practice can feel pessimistic, what I wish to leave readers of this book with is the hope that new choices are possible, that policies can be changed, that everyday life can be rebuilt away from modes of surveillance and control and toward celebrations of the complexity of lived experience. We can develop and advocate for new choices and uses of technologies that celebrate the spontaneity, richness, and chaos of everyday life rather than continuing to allow it to be governed by the potentially oppressive forces charted throughout this book. We might choose to use wearable technologies to revel in the messiness of everyday life.

The project of wearables could be to bring some chaos to life.

NOTES

INTRODUCTION: BRINGING ORDER TO LIFE

1. Dan McGuire, "Wearables Conference Explores Fringes of Fashion, Technology," *Tech*, October 21, 1997, LexisNexis Academic.

2. For more on the MIT Media Lab, see Stewart Brand, *The Media Lab: Inventing the Future at MIT* (New York: Penguin Books, 1988).

3. McGuire, "Wearable Conference Explores Fringes."

4. For an intellectual history of technological determinism, see John Durham Peters, "'You Mean My Whole Fallacy Is Wrong': On Technological Determinism," *Representations* 140 (2017): 10–26.

5. The framing of "always-on/always-on-you" comes from Sherry Turkle, "Always-on/Always-on-You: The Tethered Self," in *Handbook of Mobile Communication Studies*, ed. James E. Katz (Cambridge, MA: MIT Press, 2008), 121–138.

6. Darrell Etheringon, "Researchers Use Biometrics, Including data from the Oura Ring, to Predict COVID-19 Symptoms in Advance," *Tech Crunch*, May 28, 2020, https://techcrunch.com/2020/05/28/researchers-use-biometrics-including-data-from-the-oura-ring-to-predict-covid-19-symptoms-in-advance/. This is discussed at greater length in chapter 1.

7. Associated Press, "Connecticut Man Sentenced to 65 years for Wife's Killing in 'Fitbit Murder' Case," *NBC News*, August 18, 2022, www.nbcnews.com/news/us-news/connecticut-man-sentenced-65-years-wifes-killing-fitbit-murder-case-rcna43859.

8. James N. Gilmore, "Deathlogging: GoPros as Forensic Media in Accidental Sporting Deaths," *Convergence: The International Journal of Research into New Media Technologies* 29, no. 2 (2023): 481–495.

9. Constantine Gidaris, "Surveillance Capitalism, Datafication, and Unwaged Labour: The Rise of Wearable Fitness Devices and Interactive Life Insurance," *Surveillance & Society* 17, nos. 1–2 (2019): 132–138. This is discussed at greater length in chapter 3.

10. Stacy E. Wood, "Police Body Cameras and Professional Responsibility: Public Records and Private Records," *Preservation, Digital Technology and Culture* 46, no. 1 (2017): 41–51.

11. Ifeoma Ajunwa, Kate Crawford, and Jason Schultz, "Limitless Worker Surveillance," *California Law Review* 105 (2017): 735–776.

12. Apple, "Rise," posted April 24, 2015, YouTube video, www.youtube.com/watch?v=_APfS8aoayQ. See also James N. Gilmore, "From Ticks and Tocks to Budges and Nudges: The Smartwatch and the Haptics of Informatic Culture," *Television and New Media* 18, no. 3 (2017): 189–202.

13. Sarah Sharma, *In the Meantime: Temporality and Cultural Politics* (Durham, NC: Duke University Press, 2014).

14. Bruno Latour, "'Where Are the Missing Masses?' The Sociology of a Few Mundane Artifacts," in *Shaping Technology/Building Society: Studies in Sociotechnical Change*, ed. Wiebe E. Bijker and John Law (Cambridge, MA: MIT Press, 1992), 225–228.

15. See, for instance, Lauren Kilgour, "The Ethics of Aesthetics: Stigma, Information, and the Politics of Electronic Ankle Monitor Design," *Information Society* 36, no. 3 (2020): 131–146.

16. James N. Gilmore and Cassidy Gruber, "Wearable Witnesses: Deathlogging and Framing Wearable Technology Data in 'Fitbit Murders,'" *Mobile Media & Communication* 12, no. 1 (2024): 195–211; Lauren Smiley, "A Brutal Murder, a Wearable Witness, and an Unlikely Suspect," *Wired*, September 17, 2019, www.wired.com/story/telltale-heart-fitbit-murder/; and Jeffrey Bellin and Shevarma Pemberton, "Policing the Admissibility of Body Camera Evidence," *Fordham Law Review* 87, no. 4 (2019): 1425–1457.

17. For some examples of this recurrent theme in technology studies literature, see Ruth Schwartz Cowan, *More Work for Mother: The Ironies of Household Technology from the Open Hearth to the Microwave* (New York: Basic Books, 1985); Jacques Ellul, *The Technological Society* (New York: Vintage Books, 1964); and Langdon Winner, *The Whale and the Reactor: A Search for Limits in an Age of High Technology* (Chicago: University of Chicago Press, 1986).

18. For more on datafication, see Viktor Mayer-Schönberger and George Cukier, *Big Data: A Revolution That Will Transform How We Live, Work, and Think* (London: John Murray, 2014); and José van Dijck, "Datafication, Dataism, and Dataveillance: Big Data between Scientific Paradigm and Ideology," *Surveillance and Society* 12, no. 2 (2014): 197–208. See also Roger Clarke, "The Digital Persona and Its Application to Data Surveillance," *Information Society* 10, no. 2 (1994): 77–92; and John Cheney-Lippold, *We Are Data: Algorithms and the Making of Our Digital Selves* (New York: New York University Press, 2017).

19. See Colin Koopman, *How We Became Our Data: A Genealogy of the Informational Person* (Chicago: The University of Chicago Press, 2019).

20. For some examples, see James N. Gilmore, "Everywear: The Quantified Self and wearable Fitness Technologies," *New Media & Society* 18, no. 11 (2016): 2524–2539; Dorothea Brogard Kristensen and Minna Ruckenstein, "Co-evolving

with Self-Tracking Technologies," *New Media & Society* 20, no. 10 (2018): 3624–3640; Giovanna Mascheroni, "Datafied Childhoods: Contextualizing Datafication in Everyday Life," *Current Sociology* 68, no. 6 (2020): 798–813; Helen Kennedy, "Living with Data: Aligning Data Studies and Data Activism through a Focus on Everyday Experiences of Datafication," *Krisis: Journal for Contemporary Philosophy* 38, no. 1 (2018): 18–30; and Jathan Sadowski, "When Data Is Capital: Datafication, Accumulation, and Extraction," *Big Data & Society* 6, no. 1 (2019): 1–12.

21. Minna Ruckenstein, "Visualized and Interacted Life: Personal Analytics and Engagements with Data Doubles," *Societies* 4, no. 1 (2014): 68–84.

22. Simone Browne has understood surveillance transformations as a spectrum across history, in which old forms of discrimination are rebuilt through, now, digital tools. See Simone Browne, *Dark Matters: On the Surveillance of Blackness* (Durham, NC: Duke University Press, 2015).

23. Charles Taylor, *Modern Social Imaginaries* (Durham, NC: Duke University Press, 2004).

24. While this is far from an exhaustive list, a studies in this area include Jaimie Lee Freeman and Gina Neff, "The Challenge of Repurposed Technologies for Youth: Understanding the Unique Affordances of Digital Self-Tracking for Adolescents," *New Media & Society* 25, no. 11 (2021): 3047–3064; Jeeyun Oh and Hyunjin Kant, "User Engagement with Smart Wearables: Four Defining Factors and a Process Model," *Mobile Media & Communication* 9, no. 2 (2021): 314–335; Mark Holton, "Walking with Technology: Understanding Mobility-Technology Assemblages," *Mobilities* 14, no. 4 (2019): 435–451; and Karl-Jacob Michelson, "'Running Is My Boyfriend': Consumers' Relationships with Activities," *Journal of Services Marketing* 31, no. 1 (2017): 24–33.

25. I take my understanding of *emergence* from Raymond Williams's concept of "emergent culture," or things that are just coming into view, with forms and patterns of practice that are not yet clearly defined. Raymond Williams, *Marxism and Literature* (Oxford: Oxford University Press, 1977).

26. See, for example, Ki Joon Kim and Dong-Hee Shin, "An Acceptance Model for Smart Watches: Implications for the Adoption of Future Wearable Technology," *Internet Research* 25, no. 4 (2015): 527–541; João J. Ferreira et al., "Wearable Technology and Consumer Interaction: A Systematic Review and Research Agenda," *Computers in Human Behavior* 118 (2021): 1–10; Dawn Nafus, "The Domestication of Data: Why Embracing Digital Data Means Embracing Bigger Questions," *Ethnographic Praxis in Industry Conference Proceedings*, no. 1 (2016): 387–399; and Dawn Nafus, *Quantified: Biosensing technologies in everyday life* (Cambridge, MA: MIT Press, 2016).

27. Michael Pickering, *History, Experience, and Cultural Studies* (New York: St. Martin's Press, 1997).

28. Pickering, *History, Experience, and Cultural Studies*, 57.

29. Henri Lefebvre noted that "everyday life" really only began to emerge as a distinct space around the turn of the twentieth century; that is to say, with the arrival of modernity. In that sense, it has been part of the development of mass media,

computation, and the social sciences, all of which emerged or dominated the first part of the twentieth century. See, for some examples of this, Henri Lefebvre, *Critique of Everyday Life*, vol. 2, *Foundations for a Sociology of the Everyday* (New York: Verso, 2004). Lefebvre noted the move toward programming and control in everyday life in the 1960s in *Everyday Life in the Modern World* (New York: Routledge, 2017). The relationship between surveillance and "everyday life" broadly considered has also been an ongoing pillar of research in surveillance studies. See, for example, David Lyon, *The Culture of Surveillance: Watching as a Way of Life* (Malden, MA: Polity, 2018); Louise Eley and Ben Rampton, "Everyday Surveillance, Goffman, and Unfocused Interaction," *Surveillance & Society* 18, no. 2 (2020): 199–215; and Nicola Green and Nils Zurawski, "Surveillance and Ethnography: Researching Surveillance as Everyday Life," *Surveillance & Society* 13, no. 1 (2015): 27–43.

30. Sadowski, "When Data Is Capital"; and Lisa Gitelman, *"Raw Data" Is an Oxymoron* (Cambridge, MA: MIT Press, 2013).

31. For an example see the textbook by Jute Leskovec, Anand Rajaraman, and Jeffrey David Ullman, *Mining of Massive Datasets* (Cambridge: Cambridge University Press, 2020).

32. Jonis Toonders, "Data Is the New Oil of the Digital Economy," *Wired*, July 2014, www.wired.com/insights/2014/07/data-new-oil-digital-economy/.

33. Chris Anderson, "The End of Theory: The Data Deluges Makes the Scientific Method Obsolete," *Wired*, June 23, 2008, www.wired.com/2008/06/pb-theory/.

34. Minna Ruckenstein and Mika Pantzar, "Beyond the Quantified Self: Thematic Exploration of a Dataistic Paradigm," *New Media & Society* 19, no. 3 (2017): 401–418.

35. Luke Stark has called this the "scalable subject." Luke Stark, "Algorithmic Psychometrics and the Scalable Subject," *Social Studies of Science* 48, no. 2 (2018): 204–231. See also James N. Gilmore, "To Affinity and Beyond: Clicking as Communicative Gesture on the Experimentation Platform," *Communication, Culture, and Critique* 13 (2020): 367–383.

36. John Laidler, "High Tech Is Watching You," *Harvard Gazette*, March 4, 2019, https://news.harvard.edu/gazette/story/2019/03/harvard-professor-says-surveillance-capitalism-is-undermining-democracy/.

37. Devon Powers, *On Trend: The Business of Forecasting the Future* (Urbana-Champaign: University of Illinois Press, 2019).

38. Maurice Blanchot, "Everyday Speech," *Yale French Studies* 73 (1987): 14.

39. John Durham Peters, *The Marvelous Clouds: Toward a Philosophy of Elemental Media* (Chicago: University of Chicago Press, 2015), 11.

40. See, for example, James R. Beniger, *The Control Revolution: Technological and Economic Origins of the Information Society* (Cambridge, MA: Harvard University Press, 1986).

41. This is an argument similar to Roland Barthes's discussions of photography in *Camera Lucida: Reflections on Photography* (New York: Hill and Wang, 2010).

42. While the literature on agenda setting and framing is massive, see as one exemplar, Dietram A. Scheufele, "Framing as a Theory of Media Effects," *Journal of Communication* 49, no. 1 (1999): 103–122.

43. Peters, *Marvelous Clouds*, 15.

44. Friedrich Kittler, *Gramophone, Film, Typewriter* (Stanford, CA: Stanford University Press, 1999); and Lisa Gitelman, *Always Already New: Media, History, and the Data of Culture* (Cambridge, MA: MIT Press, 2008).

45. Turkle, "Always-on/Always-on-You."

46. Sherry Turkle, *Alone Together: Why We Expect More from Technology and Less from Each Other* (New York: Basic Books, 2012).

47. Another book that discusses wearables within the fashion industry from a technical standpoint is Jane McCann and David Bryson, *Smart Clothes and Wearable Technology* (Boca Raton, FL: CRC Press, 2009).

48. Adi Robertson, "A Wisconsin Company Will Let Employees Use Microchip Implants to Buy Snacks and Open Doors," *Verge*, July 24, 2017, www.theverge.com/2017/7/24/16019530/three-sqaure-market-implant-office-keycard-biohacking-wisconsin.

49. Susan Elizabeth Ryan, *Garments of Paradise: Wearable Discourse in the Digital Age* (Cambridge, MA: MIT Press, 2014), 238.

50. This is a point well made by Henri Lefebvre across his works on the critique of everyday life, especially *Everyday Life in the Modern World*. For some discussion of how this might be updated to the twenty-first century, see Ted Striphas, *The Late Age of Print: Everyday Book Culture from Consumerism to Control* (New York: Columbia University Press, 2009); and James N. Gilmore, "Alienating and Reorganizing Cultural Goods: Using Lefebvre's Controlled Consumption Model to Theorize Media Industry Change," *International Journal of Communication* 14 (2020): 4474–4493.

51. See, for example, Dick Hebdige, *Subculture: The Meaning of Style* (New York: Routledge, 1979); and Erving Goffman, *The Presentation of Self in Everyday Life* (New York: Anchor Books, 1959).

52. See Goffman, *Presentation of Self in Everyday Life*; and Judith Butler, *Gender Trouble: Feminism and the Subversion of Identity* (New York: Routledge, 1992).

53. For more on this distinction, see Ted Striphas, "The Visible College," *International Journal of Communication* 5 (2011): 1744–1751.

54. Wendy Hui Kyong Chun, *Updating to Remain the Same: Habitual New Media* (Cambridge, MA: MIT Press, 2016).

55. Michael Gardiner, *Critiques of Everyday Life* (New York: Routledge, 2000).

56. Lennard Davis, *Enforcing Normalcy: Disability, Deafness, and the Body* (New York: Verso, 1995), 3.

57. Sara Ahmed, *Queer Phenomenology: Orientations, Objects, Others* (Durham, NC: Duke University Press, 2006).

58. Davis, *Enforcing Normalcy*, 5.

59. Guy C. Le Masurier, Cara L. Sidman, and Charles B. Corbin, "Accumulating 10,000 Steps: Does This Meet Current Physical Activity Guidelines?" *Research Quarterly for Exercise and Sport* 74, no. 4 (2003): 389–394.

60. Harley Pasternak, "The Power of Taking 10,000 Steps (or More!) and How to Get There," *Fitbit* (blog), March 23, 2018, https://blog.fitbit.com/walking-10000-steps-a-day/.

61. Daniel Lieberman, *Exercised: Why Something We Never Evolved to Do Is Healthy and Rewarding* (New York: Pantheon, 2021).

62. I-Min Lee et. al., "Association of Step Volume and Intensity with All-Cause Mortality in Older Women," *JAMA Internal Medicine* 179, no. 8 (2019): 1105–1112. See also Amanda Mull, "What 10,000 Steps Will Really Get You," *Atlantic* May 31, 2019, www.theatlantic.com/health/archive/2019/05/10000-steps-rule/590785/.

63. Marqui Mapp, "John Hancock's Bargain: Give Us Your Data, You Pay Less in Rates," *CNBC*, April 19, 2015, www.cnbc.com/2015/04/19/john-hancocks-bargain-give-us-more-data-you-pay-less-in-rates.html.

64. There are also resonances here with Michel Foucault's conceptions of self-care and technologies of the self. See Michel Foucault, *History of Sexuality*, vol. 3, *The Care of the Self* (New York: Vintage, 1988).

65. Sarah F. Roberts-Lewis et al., "Validity of Fitbit Activity Monitoring for Adults with Progressive Muscle Diseases," *Disability and Rehabilitation* 44, no. 24 (2022): 7543–7553.

66. Exemplary work in this area includes Jonathan Sterne, *Diminished Faculties: A Political Phenomenology of Impairment* (Durham, NC: Duke University Press, 2022); Elizabeth Ellcessor, *Restricted Access: Media, Disability, and the Politics of Participation* (New York: New York University Press, 2016); and Mack Hagood, *Hush: Media and Sonic Self-Control* (Durham, NC: Duke University Press, 2019). For a comprehensive overview of this subfield, see Bill Kirkpatrick and Elizabeth Ellcessor, *Disability Media Studies* (New York: New York University Press, 2017).

67. Steve Mann, "Sousveillance," WearCam, accessed July 1, 2024, http://wearcam.org/sousveillance.htm; and Steve Mann, "'Sousveillance': Inverse Surveillance in Multimedia Imaging," *Proceedings of the 12th Annual ACM International Conference on Multimedia* (2004): 620–627. As surveillance studies researcher Torin Monahan has suggested, Mann's sousveillance does not always work in practice, noting how some of Mann's wearable camera footage involves "targeting black women in low-wage service sector jobs and thrusting his camera in their faces in the service of his critical performances." Torin Monahan, "Reckoning with COVID, Racial Violence, and the Perilous Pursuit of Transparency," *Surveillance & Society* 19, no. 1 (2021): 4.

68. James N. Gilmore, "Securing the Kids: Geofencing and Child Wearables," *Convergence* 26, nos. 5–6 (2020): 1333–1346; and Amy Adele Hasinoff, "Where Are You? Location Tracking and the Promise of Child Safety," *Television & New Media* 18, no. 6 (2017): 496–512.

69. Karen Levy, *Data Driven: Truckers, Technology, and the New Workplace Surveillance* (Princeton, NJ: Princeton University Press, 2022).

70. Hassinoff, "Where Are You?"

71. Sarah Pink and Vaike Fors, "Self-Tracking and Mobile Media: New Digital Materialities," *Mobile Media & Communication* 5, no. 3 (2017): 219–238.

72. Lyon, *Culture of Surveillance.*

73. Browne, *Dark Matter.*

74. Gilles Deleuze, "Postscript on the Societies of Control," *October* 59 (1992): 3–7.

75. Evgeny Morozov, *To Save Everything, Click Here: The Folly of Technological Solutionism* (New York: Public Affairs, 2013).

76. While chapter 5 goes into detail on this through a specific case study, see for an overview on these issues C. J. Reynolds, "Mischievous Infrastructure: Tactical Secrecy through Infrastructural Friction in Police Video Systems," *Cultural Studies* 35, nos. 4–5 (2021): 996–1019.

77. Stefania Milan, "Techno-solutionism and the Standard Human in the Making of the COVID-19 Pandemic," *Big Data & Society* 7, no. 2 (2020): 1–7.

78. For some instances of this, see Linnet Taylor, "There Is an App for That: Technological Solutionism as COVID-19 Policy in the Global North," in *The New Common: How the COVID-19 Pandemic Is Transforming Society*, ed. Emile Aarts et al. (Cham, Switzerland: Springer, 2021), 209–216; Luca Martelli, Katharine Kieslich, and Susi Geiger, "COVID-19 and Techno-solutionism: Responsibilization without Contextualization?," *Critical Public Health* 32, no. 1 (2022): 1–4; and Dillon Wamsley and Benjamin Chin-Yee, "COVID-19, Digital Health Technology and the Politics of the Unprecedented," *Big Data & Society* 8, no. 1 (2021): 1–6.

79. Langdon Winner, *Autonomous Technology: Technics-Out-of-Control as a Theme in Political Thought* (Cambridge, MA: MIT Press, 1977).

80. B. R, Cohen, "Public Thinker: Jill Lepore on the Challenge of Explaining Things," *Public Books* April 24, 2017, www.publicbooks.org/public-thinker-jill-lepore-on-the-challenge-of-explaining-things/.

81. Kilgour, "Ethics of Aesthetics."

82. Max Goetschel and Jon M. Phea, "Police Perceptions of Body-Worn Cameras," *American Journal of Criminal Justice* 42 (2017): 698–726.

83. See, for an overview of articulation, Jennifer Daryl Slack, "The Theory and Method of Articulation in Cultural Studies," in *Stuart Hall: Critical Dialogues in Cultural Studies*, ed. David Morley and Kuan-Hsing Chen (New York: Routledge, 1992), 112–127.

84. Slack, "Theory of Articulation."

85. Raymond Williams, *Culture and Society: 1780–1950* (New York: Anchor Books, 1960).

86. See Stuart Hall et al., *Policing the Crisis: Mugging, the State, and Law and Order* (New York: Palgrave, 1978). For further elaboration of "the conjuncture" as Hall conceptualized it, see Jeremy Gilbert, "This Conjuncture: For Stuart Hall," *New Formations: A Journal of Culture/Theory/Politics* 96–97 (2019): 5–37; see also Lawrence Grossberg, "Cultural Studies in Search of a Method, or Looking for

Conjunctural Analysis," *New Formations: A Journal of Culture/Theory/Politics* 96–97 (2019): 38–68.

87. Lawrence Grossberg, *Cultural Studies in the Future Tense* (Durham, NC: Duke University Press, 2010).

88. Peters, "'My Fallacy Is Wrong.'"

89. Predominantly, Peters considers Friedrich Kittler's opening proclamation in his book *Gramophone, Film, Typewriter*, that "media determine our situation." He also cites Langdon Winner, Alfred Chandler, David Noble, Raymond Williams, E. P. Thompson, and others as trying to nuance technological determinism into a cultural politics of technology that recognizes the role technologies do play in shaping social conditions.

90. Winner, *Whale and the Reactor*, 26.

91. Norman Fairclough, *Critical Discourse Analysis: The Critical Study of Language* (New York: Routledge, 1995).

92. This idea of mapping has also recently been used by Kate Crawford in *Atlas of AI* to describe different domains of artificial intelligence. Kate Crawford, *Atlas of AI: Power, Politics, and the Planetary Costs of Artificial Intelligence* (New Haven, CT: Yale University Press, 2021).

CHAPTER 1: HEALTH, OR: BRINGING THE HOSPITAL TO THE WRIST

1. Heather Landy, "See What Happened to Twitter CEO Jack Dorsey's Heart Rate as He Testified to Congress," *Quartz*, September 5, 2018, https://qz.com/work/1380590/twitter-ceo-jack-dorseys-heart-rate-rose-as-he-testified-to-congress.

2. For instance, Maarten Falter et al., "Accuracy of Apple Watch Measurements for Heart Rate and Energy Expenditure in Patients with Cardiovascular Disease: Cross-Sectional Study," *JMIR Mhealth Uhealth* 7, no. 3 (2019): e11889; and Matthew P. Wallet et al., "Accuracy of Heart Rate Watches: Implications for weight management," *PLoS One* 11, no. 5 (2016): e0154420.

3. Ulrich Beck, *Risk Society: Towards a New Modernity* (London: Sage Publications, 1992).

4. "Monitor Your Heart Rate with Apple Watch," Apple, accessed May 20, 2023, https://support.apple.com/en-us/HT204666.

5. For one example of this, see Ezekiel Emanuel, Aaron Glickman, and David Johnson, "Measuring the Burden of Health Care Costs on US Families: The Affordability Index," *JAMA* 318, no. 19 (2017): 1863–1864.

6. For some discussion of this "democratization" effect from the point of view of clinical research, see Serge Korjian and C. Michael Gibson, "Digital Technologies and the Democratization of Clinical Research: Social Media, Wearables, and Artificial Intelligence," *Contemporary Clinical Trials* 117 (2022): 106767.

7. See, for example, Natasha Dow Schüll, "Data for Life: Wearable Technology and the Design of Self-Care," *BioSocieties* 11, no. 3 (2016): 317–333; and Artistes

Fotopoulou and Kate O'Riordan, "Training to Self-Care: Fitness Tracking, Biopedagogy and the Healthy Consumer," *Health Sociology Review* 26, no. 1 (2017): 54–68. This discussion is continued in chapter 3 and engages more directly with Michel Foucault's historical conceptualizations of self-care.

8. Gary Wolf, "The Data-Driven Life," *New York Times*, April 28, 2010, www.nytimes.com/2010/05/02/magazine/02self-measurement-t.html.

9. Melanie Swan, "The Quantified Self: Fundamental Disruption in Big Data Science and Biological Discovery," *Big Data* 1, no. 2 (2013): 85–99; and Eloisa Prasopolou, "A Half-Moon on My Skin: A Memoir on Life with an Activity Tracker," *European Journal of Information Systems* 26, no. 3 (2017): 287–297.

10. Gary Wolf, "Know Thyself: Tracking Every Facet of Life, from Sleep to Mood to Pain, 24/7/365," *Wired*, June 22, 2009, www.wired.com/2009/06/lbnp-knowthyself/.

11. James Gleick, *The Information: A History, a Theory, a Flood* (New York: Pantheon Books, 2011).

12. I use the phrase "social laboratory" as a play on the autonomous Marxist notion of the "social factory," in which processes of work extend into the social sphere. Here, processes of scientific experimentation become the logic through which to understand everyday life and social negotiation. See Rosalind Gill and Andy Pratt, "In the Social Factory? Immaterial Labour, Precariousness and Cultural Work," *Theory, Culture, and Society* 25, nos. 7–8 (2008): 1–30. For more research on the quantified self movement, see Deborah Lupton, *The Quantified Self* (Malden, MA: Polity, 2016). See also Gina Neff and Dawn Nafus, *Self-Tracking* (Cambridge, MA: MIT Press, 2016).

13. This is, of course, nothing new. Historian J. Rosser Matthews has demonstrated how medical practitioners in Europe as early as the 1780s were embroiled in debates about the capacity to reduce knowledge about bodies to statistics. J. Rosser Matthews, *Quantification and the Quest for Medical Certainty* (Princeton, NJ: Princeton University Press, 1995).

14. Patti Shih et al., "Direct-to-Consumer Detection of Atrial Fibrillation in a Smartwatch Electrocardiogram: Medical Overuse, Medical Inaction and the Experience of Consumers," *Social Science & Medicine* 303 (2022): 114954.

15. Eric Topol, *The Patient Will See You Now: The Future of Medicine Is in Your Hands* (New York: Basic Books, 2015).

16. Michel Foucault, *The Birth of the Clinic: An Archaeology of Medical Perception* (London: Tavistock, 1973).

17. For one of the earliest developments of this term that I am aware of, see Kevin D. Haggerty and Richard V. Ericson, "The Surveillant Assemblage," *British Journal of Sociology* 51, no. 4 (2000): 605–622.

18. Minna Ruckenstein, "Visualized and Interacted Life: Personal Analytics and Engagements with Data Doubles," *Societies* 4 (2014): 78.

19. For some examples of this, see John Cheney-Lippold, *We Are Data: Algorithms and the Making of Our Digital Selves* (New York: New York University Press, 2017); David Lyon, "Everyday Surveillance: Personal Data and Social

Classifications," *Information, Communication, and Society* 5, no. 2 (2002): 242–257; Blake Hallinan and Ted Striphas, "Recommended for You: The Netflix Prize and the Algorithmic Production of Culture," *New Media & Society* 18, no. 1 (2016): 117–137; and James N. Gilmore, "To Affinity and Beyond: Clicking as Communicative Gesture on the Experimentation Platform," *Communication, Culture, and Critique* 13 (2020): 367–383.

20. For some examples that position the politics of biometrics, see Kelly A. Gates, *Our Biometric Future: Facial Recognition Technology and the Culture of Surveillance* (New York: New York University Press, 2011); and Btihaj Ajana, *Governing through Biometrics: The Biopolitics of Identity* (New York: Palgrave Macmillan, 2013), 3.

21. Ajana, *Governing through Biometrics*, 21.

22. For more on the tense interplays between democracy and surveillance, see Kevin Haggerty and Minas Samatras, *Surveillance and Democracy* (New York: Routledge, 2010).

23. For some examples of this, see Geoffrey C. Bunn, *The Truth Machine: A Social History of the Lie Detector* (Baltimore, MD: Johns Hopkins University Press, 2012); and Jonathan Sterne, *The Audible Past: Cultural Origins of Sound Reproduction* (Durham, NC: Duke University Press, 2003).

24. "How Does My Tracker Count Steps?" Fitbit, October 28, 2016, https://help.fitbit.com/articles/en_US/Help_article/1143.

25. There are broad connections that might be made here to theories of simulation, such as Jean Baudrillard, *Simulacra and Simulation*, trans. Sheila Faria Glaser (Ann Arbor: University of Michigan Press, 1994).

26. For some discussion of this idea in relation to virtual reality space, see Ken Hillis, *Digital Sensations: Space, Identity, and Embodiment in Virtual Reality* (Minneapolis: University of Minnesota Press, 1999).

27. Sarah F. Roberts-Lewis et al., "Validity of Fitbit Activity Monitoring for Adults with Progressive Muscle Disease," *Disability and Rehabilitation* 44, no. 24 (2022): 7543–7553.

28. "Wheelchair User: Push/Step Counting," Fitbit Community, accessed December 1, 2019, https://community.fitbit.com/t5/Other-Inspire-Trackers/Wheelchair-User-push-step-counting/td-p/3337961.

29. Here, one might think about the notion of "the limit" in calculus, which posits a value that is infinitely approachable, but never attainable. Gilles Deleuze explores this at length in the context of repetition and replication in *Difference and Repetition* (New York: Columbia University Press, 1995).

30. Much of this information is designed to be "glanced at" rather than thoroughly considered. This concept of "the glance" is discussed in more detail in James N. Gilmore, "From Ticks and Tocks to Budges and Nudges: The Smartwatch and the Haptics of Informatic Culture," *Television and New Media* 18, no. 3 (2017): 189–202. See also Diana Zulli, "Capitalizing on the Look: Insights into the Glance, Attention Economy, and Instagram," *Critical Studies in Media Communication* 35, no. 2 (2018): 137–150.

31. Rebecca Kizer, "Fitbit Facing Lawsuit Due to Inaccuracy," *Ball State Daily*, May 30, 2016, www.ballstatedaily.com/article/2016/05/news-fitbit-lawsuit.

32. Kayleena Markortoff, "Study Claims Fitbit Trackers Are 'Highly Inaccurate,'" *CNBC*, May 23, 2016, www.cnbc.com/2016/05/23/study-shows-fitbit-trackers-highly-inaccurate.html.

33. Edward Jo and Brett A. Dolezal, "Validation of the Fitbit Surge and Charge HR Fitness Trackers," 2016, www.lieffcabraser.com/pdf/Fitbit_Validation_Study.pdf.

34. Kayleena Markortoff, "Fitbit Trackers 'Highly Inaccurate.'"

35. "Fitbit Charge HR TV Commercial, 'Know Your Heart,'" iSpot, accessed May 30, 2022, www.ispot.tv/ad/7QpS/fitbit-charge-hr-know-your-heart; and "Fitbit Charge 2—Big Day," posted April 5, 2017, YouTube video, www.youtube.com/watch?v=bW5ihFg-5iQ.

36. Brian Fung, "Is Your Fitbit Wrong? One Woman Argued Hers Was—and Almost Ended Up in a Legal No-Man's Land," *Washington Post*, August 2, 2018, www.washingtonpost.com/technology/2018/08/02/is-your-fitbit-wrong-one-woman-argued-it-was-almost-ended-up-legal-no-mans-land/.

37. "Consumers File Class Action Against Fitbit, Inc. Alleging That Fitbit Charge HR and Surge Heart Rate Monitors Do Not Accurately Track Users' Heart Rates," *Business Wire*, January 5, 2016, www.businesswire.com/news/home/20160105006634/en/Consumers-File-Class-Action-Against-Fitbit-Inc.-Alleging-That-Fitbit-Charge-HR-and-Surge-Heart-Rate-Monitors-Do-Not-Accurately-Track-Users'-Heart-Rates. See Lauren Goode, "Fitbit Hit with Class-Action Suit over Inaccurate Heart Rate Monitoring," *Verge*, January 6, 2016, www.theverge.com/2016/1/6/10724270/fitbit-lawsuit-charge-hr-surge-incomplete-heart-rate-tracking.

38. Kate McLellan et al. v. Fitbit Inc., "Order Re Motion to Dismiss,." June 5, 2018, www.lieffcabraser.com/pdf/Fitbit-060518-Order-re-Motion-to-Dismiss.pdf.

39. Fung, "Is Your Fitbit Wrong?."

40. Kate McLellan et al. v. Fitbit, Inc., "Order Re Arbitration Proceedings," July 24, 2018, www.govinfo.gov/content/pkg/USCOURTS-cand-3_16-cv-00036/pdf/USCOURTS-cand-3_16-cv-00036-5.pdf.

41. Dan Bouk, *How Our Days Became Numbered: Risk and the Rise of the Statistical Individual* (Chicago: University of Chicago Press, 2015).

42. Beck, *Risk Society*, 19–20.

43. Nikolas Rose, *Powers of Freedom: Reframing Political Thought* (Cambridge: Cambridge University Press, 1999). For a more recent discussion of the utility of "responsibilization," see Jarkko Pyysiäinen, Darren Halpin, and Andrew Guilfoyle, "Neoliberal Governance and 'Responsibilization' of Agents: Reassessing the Mechanisms of Responsibility-Shift in Neoliberal Discursive Environments," *Distinktion: Journal of Social Theory* 18, no. 2 (2017): 215–235.

44. "Fitbit Corporate Wellness Solutions," Fitbit, accessed September 8, 2019, www.fitbit.com/en-ca/product/corporate-solutions; and Jonah Comstock, "Fitbit CEO Hints at Expanding Healthcare Strategy, FDA-Cleared Devices," *Mobi Health News*, May 5, 2016, www.mobihealthnews.com/content/fitbit-ceo-hints-expanding-healthcare-strategy-fda-cleared-devices.

45. Lizzy Gurdus, "United Healthcare and Fitbit to pay Users up to $1,500 to Use Devices, Fitbit Co-founder Says," *CNBC*, January 5, 2017, www.cnbc.com/2017/01/05/unitedhealthcare-and-fitbit-to-pay-users-up-to-1500-to-use-devices.html.

46. "Fitbit Launches Fitbit Care, a Powerful New Enterprise Health Platform for Wellness and Prevention and Disease Management," Fitbit, September 18, 2018, https://investor.fitbit.com/press/press-releases/press-release-details/2018/Fitbit-Launches-Fitbit-Care-A-Powerful-New-Enterprise-Health-Platform-for-Wellness-and-Prevention-and-Disease-Management/default.aspx.

47. "Introducing Fitbit Care," Fitbit Health Solutions, accessed September 1, 2019, healthsolutions.fitbit.com/fitbit-care/#tools-to-manage-and-measure-your-program.

48. "Introducing Fitbit Care."

49. "John Hancock Introduces a Whole New Approach to Life Insurance in the U.S. That Rewards Customers for Healthy Living," *PR Newswire*, April 8, 2015, www.prnewswire.com/news-releases/john-hancock-introduces-a-whole-new-approach-to-life-insurance-in-the-us-that-rewards-customers-for-healthy-living-300062461.html.

50. Suzanne Barlyn, "Strap on the Fitbit: John Hancock to Sell Only Interactive Life Insurance," *Reuters*, September 19, 2018, www.reuters.com/article/us-manulife-financi-john-hancock-lifeins/strap-on-the-fitbit-john-hancock-to-sell-only-interactive-life-insurance-idUSKCN1LZ1WL.

51. "Get your Apple Watch with John Hancock Vitality," John Hancock, accessed September 7, 2019, www.johnhancockinsurance.com/vitality-program/apple-watch.html.

52. This extends some of the ways Lana Swartz has conceptualized money as a medium. See Lana Swartz, *New Money: How Payment Became Social Media* (New Haven, CT: Yale University Press, 2020).

53. John Hancock/Vitality, home page, accessed September 6, 2019, https://termlife.johnhancockinsurance.com/mvt/vitality-life-quote-combined.

54. "John Hancock Leaves Traditional Life Insurance Model Behind to Incentivize Longer, Healthier Lives," John Hancock, September 19, 2018, www.johnhancock.com/news/insurance/2018/09/john-hancock-leaves-traditional-life-insurance-model-behind-to-incentivize-longer--healthier-lives.html.

55. This idea was popularized in Richard H. Thaler and Cass R. Sunstein, *Nudge: Improving Decisions about Health, Wealth, and Happiness* (New York: Penguin Books, 2009).

56. "Hancock Leaves Traditional Model."

57. Paul Sullivan, "Life Insurance Offering More Incentive to Live Longer," *New York Times*, September 19, 2018, nytimes.com/2018/09/19/your-money/john-hancock-vitality-life-insurance.html.

58. Daniel J. Solove, "'I've Got Nothing to Hide' and Other Misunderstandings of Privacy," *San Diego Law Review* 44 (2007): 745–772.

59. BMI is historically considered an inaccurate way to assess bodily health, serving instead as a reductive variable. For example, see Lee F. Monaghan, Rachel

Colls, and Bethan Evans, "Obesity discourse and fat politics: research, critique and interventions," *Critical Public Health* 23, no.3 (2013): 249–262.

60. "Earn Vitality Points for Healthy Living," John Hancock/Vitality, accessed September 5, 2019, https://jh1.jhlifeinsurance.com/jhl-ext-templating/filedetail?vgnextoid=83b97a0fbe21c410VgnVCM1000003e86fa0aRCRD&siteName=JHSalesNet.

61. This gold version became functionally obsolete by 2018. See Avery Hartmans, "The $10,000 Apple Watch Will Stop Getting Major Software Updates from Apple Starting This Fall," *Business Insider*, June 4, 2018, www.businessinsider.com/apple-watch-gold-edition-will-not-work-with-watchos-5-2018-6.

62. Gilmore, "From Ticks and Tocks."

63. Lauren Goode, "How Does the Apple Watch Stack Up as a Health-and-Fitness Tracker?," *Vox*, April 20, 2015, www.vox.com/2015/4/20/11561634/health-and-fitness-on-apple-watch-a-solid-start-with-limitations.

64. Gilmore, "From Ticks and Tocks."

65. Angela Chen, "What the Apple Watch's FDA Clearance Actually Means," *Verge*, September 13, 2018, www.theverge.com/2018/9/13/17855006/apple-watch-series-4-ekg-fda-approved-vs-cleared-meaning-safe.

66. Food and Drug Administration, "De Novo Classification Request for ECG App," August 14, 2018, www.accessdata.fda.gov/cdrh_docs/reviews/DEN180044.pdf.

67. Stanford Medicine, "Through Apple Heart Study, Stanford Medicine Researchers Show Wearable Technology Can Help Detect Atrial Fibrillation," Stanford Medicine News Center, November 13, 2019, https://med.stanford.edu/news/all-news/2019/11/through-apple-heart-study--stanford-medicine-researchers-show-we.html.

68. Stanford Medicine, "Through Apple Heart Study." For the complete study, see Marco V, Perez et al., "Large-Scale Assessment of a Smartwatch to Identify Atrial Fibrillation," *New England Journal of Medicine* 381 (2019): 1909–1917.

69. "Stanford Medicine Announces Results of Unprecedented Apple Heart Study," Apple, March 16, 2019, www.apple.com/newsroom/2019/03/stanford-medicine-announces-results-of-unprecedented-apple-heart-study/.

70. John Koetsier, "Apple Heart Study: What Stanford Medicine Learned from 400,000 Apple Watch Owners," *Forbes*, March 18, 2019, www.forbes.com/sites/johnkoetsier/2019/03/18/apple-heart-study-what-stanford-medicine-learned-from-400000-apple-watch-owners/?sh=48f7c5ee2d20.

71. Chris Welch, "Apple and Stanford's Apple Watch Study Identified Irregular Heartbeats in over 2,000 Patients," *Verge*, March 16, 2019, www.theverge.com/2019/3/16/18268559/stanford-apple-heart-study-results-apple-watch; and Christina Farr, "Apple Heart Study Shows a Lot of Promise for Digital Health, but Cardiologists Still Have Questions," *CNBC*, November 13, 2019, www.cnbc.com/2019/11/13/apple-releases-heart-study-results.html.

72. Apple, "Apple Watch—Dear Apple—Apple," posted September 12, 2017, YouTube video, www.youtube.com/watch?v=N-x8Ik9G5Dg.

73. Julie Passanante Elman, "'Find Your Fit': Wearable Technology and the Cultural Politics of Disability," *New Media and Society* 20, no. 10 (2018): 3760–3777.

74. Apple, "Apple Watch/Dear Apple/Apple," posted September 7, 2022, YouTube video, www.youtube.com/watch?v=fOHj5kGU4fY.

75. Lauren Goode, "Can a Wearable Detect Covid-19 before Symptoms Appear?," *Wired*, April 14, 2020, www.wired.com/story/wearable-covid-19-symptoms-research/.

76. Shams Chariana, "Inside the NBA Bubble: Details from the NBPA Memo Obtained by The Athletic," *Athletic*, June 16, 2020, https://theathletic.com/1876737/2020/06/16/inside-the-nba-bubble-details-from-nbpa-memo-obtained-by-the-athletic/.

77. Benjamin L. Smarr et al., "Feasibility of Continuous Fever Monitoring Using Wearable Devices," *Scientific Reports* 10 (2020): 21640.

78. Geoffrey A. Fowler, "Wearable Tech Can Spot Coronavirus Symptoms before You Even Realize You're Sick," *Washington Post*, May 28, 2020, www.washingtonpost.com/technology/2020/05/28/wearable-coronavirus-detect/.

79. Siva Vaidhyanathan traces similar ways Google positioned itself as a means to address civic failures in the first decade of its operations in *The Googlization of Everything (and Why We Should Worry)* (Berkeley: University of California Press, 2011).

80. For an overview of some various contexts of this prediction, see Adrian Mackenzie, "The Production of Prediction: What Does Machine Learning Want?," *European Journal of Cultural Studies* 18, nos. 4–5 (2015): 429–455; Sarah Brayne, *Predict and Surveil: Data, Discretion, and the Future of Policing* (New York: Oxford University Press, 2020); Virginia Eubanks, *Automating Inequality: How High-Tech Tools Profile, Police, and Punish the Poor* (New York: St. Martin's Press, 2019); and Mark Andrejevic, *Automated Media* (New York: Routledge, 2019).

81. "Technology in the Oura Ring," Oura, accessed July 28, 2020, https://ouraring.com/blog/ring-technology/.

82. Oura, "Readiness: Your Complete Guide," Oura Ring, 2020, https://ouraring.com/readiness-score.

83. Oura, "The Oura Difference," Oura Ring, 2020, https://ouraring.com/the-Oura-difference.

84. Oura, "Terms of Use," Oura Ring, 2018, https://ouraring.com/terms-and-conditions.

85. For more on the idea of forecasting and trends, see Devon Powers, *On Trend: The Business of Forecasting the Future* (Champaign: University of Illinois Press, 2019).

86. Rockefeller Neuroscience Institute, "Understanding the Spread: Protecting Our Health and Economy," 2020, http://wvumedicine.org/RNI/COVID19.

87. Rockefeller Neuroscience Institute, "Understanding the Spread."

88. "WVU Rockefeller Neuroscience Institute Announces Capability to Predict COVID-19 Related Symptoms up to Three Days in Advance," West Virginia University School of Medicine, May 28, 2020, https://medicine.hsc.wvu.edu/news

/story?headline=wvu-rockefeller-neuroscience-institute-announces-capability-to-predict-covid-19-related-symptoms-up-.

89. "Capability to Predict COVID-19."

90. Smarr et al., "Feasibility of Fever Monitoring," 7.

91. Christine Fisher, "Researchers Say Oura Rings Can Predict COVID-19 Symptoms Three Days Early," *Engadget*, June 1, 2020, www.engadget.com/west-virginia-university-oura-ring-covid-19-symptoms-003239603.html.

92. Samantha Previte, "NBA to Use 'Smart Rings,' Big Data to Fight Coronavirus in Disney Bubble," *New York Post*, June 19, 2020, https://nypost.com/2020/06/19/nba-to-use-smart-rings-to-detect-coronavirus-within-bubble/.a

93. Darrell Etherington, "Researchers Use Biometrics, Including Data from the Oura Ring, to Predict COVID-19 Symptoms in Advance," *TechCrunch*, May 28, 2020, https://techcrunch.com/2020/05/28/researchers-use-biometrics-including-data-from-the-oura-ring-to-predict-covid-19-symptoms-in-advance/.

94. Amy Maxmen and Jeff Tollefson, "Two Decades of Pandemic War Games Failed to Account for Donald Trump," *Nature* 584, no. 7819 (2020): 26–29.

95. Emily Abbate, "Here's How the NBA's Coronavirus-Fighting Ring Might Help," *GQ*, June 17, 2020, www.gq.com/story/oura-ring-nba.

96. Corinne Reichert, "NBA Players Could Wear Smart Ring to Track COVID-19 Symptoms as Season Resumes," *CNET*, June 22, 2020, www.cnet.com/tech/mobile/nba-players-could-wear-a-smart-ring-to-track-covid-19-symptoms-as-season-resumes-at-disney-world/.

97. Aaron Mak, "What the NBA's $300 COVID-Detecting Rings Can Actually Accomplish," *Slate*, June 22, 2020, https://slate.com/technology/2020/06/nba-coronavirus-oura-ring-orlando.html.

98. Ben Pickman, "The Story behind the Ring That Is Key to the NBA's Restart," *Sports Illustrated*, July 1, 2020, www.si.com/nba/2020/07/01/oura-ring-nba-restart-orlando-coronavirus.

99. Chase DiBenedetto, "The Apple Watch's Blood Oxygen Measurement Might Be Guilty of Racial Bias," *Mashable*, December 28, 2022, https://mashable.com/article/apple-watch-oximeter-racial-bias-lawsuit.

100. For more information on this, see Lisa Eadicicco, "Smartwatches Have Measured Blood Oxygen for Years: But Is This Useful?," *CNET*, June 16, 2022, www.cnet.com/tech/mobile/smartwatches-have-measured-blood-oxygen-for-years-but-is-it-useful/. For a more formal academic study, see Edward D. Chan, Michael M. Chan, and Mallory M. Chan, "Pulse Oximetry: Understanding Its Basic Principles Facilitates Appreciation of Its Limitations," *Respiratory Medicine* 107, no. 6 (2013): 789–799.

101. Nicole Wetsman, "Light Sensors on Wearables Struggle with Dark Skin and Obesity," *Verge*, January 21, 2022, www.theverge.com/2022/1/21/22893133/apple-fitbit-heart-rate-sensor-skin-tone-obesity.

102. Samar Baja, "Racial Bias Is Built into the Design of Pulse Oximeters," *Washington Post*, July 27, 2022, www.washingtonpost.com/made-by-history/2022/07/27/racial-bias-is-built-into-design-pulse-oximeters/.

103. Alex Morales v. Apple, Inc., "Class Action Complaint" (filed Dec. 24, 2022), https://storage.courtlistener.com/recap/gov.uscourts.nysd.591590/gov.uscourts.nysd.591590.1.0.pdf.

104. For some theoretical consideration of this, see Langdon Winner, *The Whale and the Reactor: A Search for Limits in an Age of High Technology* (Chicago: University of Chicago Press, 1988).

105. Morales v. Apple Inc., 22-CV-10872 (JSR) (S.D.N.Y. Aug. 29, 2023), https://casetext.com/case/morales-v-apple-inc-2.

CHAPTER 2: ACCESSIBILITY, OR: PERSONALIZATION AND THE PROMOTION OF HEARABLES

1. Tia DeNora, "Music as a Technology of the Self," *Poetics* 27, no. 1 (1999): 31–56; and Michael Bull, *Sounding Out the City: Personal Stereos and the Management of Everyday Life* (New York: Berg, 2000).

2. Peter Nagy and Gina Neff, "Imagined Affordance: Reconstructing a Keyword for Communication Theory," *Social Media + Society* (July–December 2015): 1–9.

3. For example, Blake Hallinan and Ted Striphas, "Recommended for You: The Netflix Prizes and the Algorithmic Production of Culture," *New Media & Society* 18, no. 1 (2017): 117–137.

4. Nick Hunn, "Hearables—the New Wearables," *Creative Connectivity* (blog), April 3, 2014, www.nickhunn.com/hearables-the-new-wearables/.

5. Jacob Kastrenakes, "The Biggest Winner from Removing the Headphone Jack Is Apple," *Verge,* September 8, 2016, www.theverge.com/2016/9/8/12839758/apple-is-biggest-winner-from-killing-headphone-jack.

6. Lauren Alix Brown, "AirPods Could Revolutionize What It Means to Be Hard of Hearing," *Quartz*, July 7, 2018, https://qz.com/1323215/apples-airpods-and-live-listen-are-a-revolution-for-the-hearing-impaired/.

7. Gerard Goggin, "Disability and Haptic Mobile Media," *New Media & Society* 19, no. 10 (2017): 1563–1580.

8. Mack Hagood, *Hush: Media and Sonic Self-Control* (Durham, NC: Duke University Press, 2019).

9. Brian Taylor, "Hearables," *Audiology Today* 27, no. 6 (2015): 22–30. See also "Regulatory Requirements for Hearing Aid Devices and Personal Sound Amplification Products—Draft Guidance for Industry and Food and Drug Administration Staff," US Food and Drug Administration, November 7, 2013, www.fda.gov/MedicalDevices/ucm373461.htm.

10. Taylor, "Hearables."

11. Mack Hagood, "Disability and Biomediation: Tinnitus as Phantom Disability," in *Disability Media Studies*, ed. Elizabeth Ellcessor and Bill Kirkpatrick (New York: New York University Press, 2018), 311–329.

12. Hagood, "Disability and Biomediation," 312.

13. Hagood, "Disability and Biomediation," 318. See also Eugene Thacker, "Biomedia," in *Critical Terms for Media Studies*, ed. W. J. T. Mitchell and Miriam B. N. Hansen (Chicago: University of Chicago Press, 2010): 117–130.

14. Albert Bregman, *Auditory Scene Analysis: The Perceptual Organization of Sound* (Cambridge, MA: MIT Press, 1990).

15. Veit Erlmann, "But What of the Ethnographic Ear? Anthropology, Sound, and the Senses," in *Hearing Cultures: Essays on Sound, Listening, and Modernity*, ed. Veit Erlmann (New York: Bloomsbury, 2004): 18.

16. Indeed, communication scholar Jonathan Sterne's histories of sound technology and media historian Friedrich Kittler's writing on systems of knowledge production about bodily senses demonstrate that hearing systems have long been the focus of such debates and concern. Jonathan Sterne, *The Audible Past: Cultural Origins of Sound Reproduction* (Durham, NC: Duke University Press, 2003); Jonathan Sterne, *MP3: The Meaning of a Format* (Durham, NC: Duke University Press, 2012); and Friedrich Kittler, *Discourse Networks, 1800/1900* (Stanford, CA: Stanford University Press, 1992).

17. Horace Newcomb and Paul M. Hirsch, "Television as a Cultural Forum: Implications for Research," *Quarterly Review of Film and Video* 8, no. 3 (1983): 45–55.

18. Donna Kornhaber, "From Posthuman to Postcinema: Crises of Subjecthood and Representation in *Her*," *Cinema Journal* 56, no. 4 (2017): 3–25; Eva-Lynn Jagoe, "Depersonalized Intimacies: The Cases of Sherry Turkle and Spike Jonze," *ESC* 42, nos. 1–2 (2016): 155–173; and Donya Maguire, "*Affairs of the Phone*: Indiewood, a Bespoke Future, and Virtual Love in Spike Jonze's *Her* (2013)," *Film Matters* (Winter 2016): 49–53.

19. For more on the history of developing Siri, see Ed Finn, *What Algorithms Want: Imagination in the Age of Computing* (Cambridge, MA: MIT Press, 2018). For discussion of other voice assistants in disability studies, see Meryl Alper, *Giving Voice: Mobile Communication, Disability, and Inequality* (Cambridge, MA: MIT Press, 2016).

20. Heather Suzanne Woods, "Asking More of Siri and Alexa: Feminine Persona in Service of Surveillance Capitalism," *Critical Studies in Media Communication* 35, no. 4 (2018): 334–349.

21. It is worth pointing out that "intuition" is a long-standing, stereotypically female trait, in contrast to stereotypical maleness that favors rationality and empiricism. For more on conceptions of prediction and intuition, see Jonathan Cohn, *The Burden of Choice: Recommendations, Subversion, and Algorithmic Culture* (New Brunswick, NJ: Rutgers University Press, 2019).

22. In this way, Samantha demonstrates some of the protocols Adrian Mackenzie traces for machine learning; namely, how data inputs help to shape the ways machine learning algorithms come to "learn about" the work and analysis they perform. See Adrian Mackenzie, "The Production of Prediction: What Does Machine Learning Want?" *European Journal of Cultural Studies* 18, no. 4 (2015): 429–445.

23. The gendering of assistant technologies as female—and the gender dynamics of office assistants—has been explored in a number of works, including Kathleen F.

McConnell, "The Profound Sound of Ernest Hemingway's Typist," *Communication and Critical Cultural Studies* 5 (2008): 325–343.

24. Giorgio Agamben, *What Is an Apparatus?* (Stanford, CA: Stanford University Press, 2009), 14. Agamben's use of "apparatus" builds off Michel Foucault's concept of the *dispositif*, discussed later in this section. The word *apparatus* has had something of a troubled history in media studies, largely because of its association with the apparatus theory of the 1970s. For Jean-Louis Baudry, the entirety of the cinema was an apparatus that worked as a confining, organizing totality. It performed the kind of controlling and intercepting work Agamben discusses, but Baudry was more convinced of cinema's deterministic function. It was not so much a question of cinema's *capacity* to control, but rather of understanding the *degree to which* its apparatus was controlling spectators. Jean-Louis Baudry, "Ideological Effects of the Basic Cinematographic Apparatus," in *Film Theory and Criticism*, 6th ed., ed. Leo Braudry and Marshall Cohen (New York: Oxford University Press, 2004), 355–365.

25. *Oxford English Dictionary*, s.v. "apparatus," accessed November 5, 2022, https://en.oxforddictionaries.com/definition/apparatus.

26. This builds on Louis Althusser's "Ideological State Apparatuses," a concept he develops to describe sites and practices that encourage the reproduction of dominant modes of thinking, such as schoolwork. That is to say, places like schools and offices "train" humans to perform in ways that ensure the ongoing reproduction of an existing social or economic order, such as wage-based capitalism. These apparatuses may not determine modes of thought, conduct, and expression, but they can "make people ready for" the conditions of reproducing the world and living within the dominant social formations. Louis Althusser, *The Reproduction of Capitalism* (New York: Verso, 2016). Agamben builds much of his understanding of the concept from theorist and historian Michel Foucault, who developed the concept *dispositif* as another way to describe apparatuses. For Foucault, a *dispositif* is "a thoroughly heterogenous ensemble consisting of discourses, institutions, architectural forms, regulatory decisions, administrative measures, scientific statements, philosophical, moral, and philanthropic positions—in short, the said and the unsaid." Michel Foucault, *Power/Knowledge: Selected Interviews and Other Writings, 1972–1977* (New York: Pantheon, 1980), 184.

27. For more on how deference and bodily management operate together, see John F. Kasson, *Rudeness and Civility: Manners in 19th Century Urban America* (New York: Hill and Wang, 1990).

28. Adrian Mackenzie, *Machine Learners: Archaeology of a Data Practice* (Cambridge, MA: MIT Press, 2017).

29. This is similar to the occasional public claims that an artificial intelligence system has become "sentient" because it has been trained to mimic human interactions. See Simone Natale, *Deceitful Media: Artificial Intelligence and Social Life after the Turing Test* (New York: Oxford University Press, 2021).

30. Eli Pariser, *The Filter Bubble: What the Internet Is Hiding from You* (New York: Penguin Books, 2012); and Joseph Turow, *The Daily You: How the New Advertising Industry Is Defining Your Identity and Your Worth* (New Haven, CT: Yale

University Press, 2013). These authors focus largely on the shifting nature of advertising and news stories, suggesting that there are ideological consequences for isolating people from the experience of mass culture.

31. This is similar to the bind of personalization that Jonathan Cohn has traced in his work, in which users seem to both want more choices and demand that choices be structured. See Cohn, *Burden of Choice*.

32. Mack Hagood, *Hush*.

33. Anne Eisenberg, "The Hearing Aid as Fashion Statement," *New York Times*, September 24, 2006, www.nytimes.com/2006/09/24/business/yourmoney/24novel.html. For more on the idea of stigma and these devices, see Margaret Wallhagen, "The Stigma of Hearing Loss," *Gerontologist* 50, no. 1 (2010): 66–75. These concerns date back to the proliferation of electric hearing aids in the first decades of the twentieth century, when advertisements would show wearers how to properly strap on and conceal the device's wires. See Mara Mills, "Hearing Aids and the History of Electronics Miniaturization," *IEEE Annals of the History of Computing* 33, no. 2 (2011): 24–44.

34. Joshua Gunn and Mirko M. Hall, "Stick It in Your Ear: The Psychodynamics of iPod Enjoyment," *Communication and Critical/Cultural Studies* 5, no. 2 (2008): 135–157. This has resonances with Thorsten Veblen's idea of conspicuous consumption, wherein affluent purchasers were compelled to constantly purchase and display new goods to demonstrate they were able to buy them. Thorstein Veblen, *Theory of the Leisure Class* (1899; New York: Dover Print, 1994).

35. Ashley Carman et al., "Are AirPods Fashionable?," *Verge*, January 25, 2017, www.theverge.com/circuitbreaker/2017/1/25/14384112/apple-airpods-fashion-style-wireless-earbuds.

36. Farhad Manjoo, "Soundhawk Smart Listening System: A Hearing Helper," *New York Times*, November 19, 2014, www.nytimes.com/2014/11/20/technology/personaltech/soundhawk-smart-listening-system-a-hearing-helper.html.

37. Margaret Rhodes, "A Sleek New Hearing Aid That Solves a Nagging Problem," *Wired*, June 24, 2014, www.wired.com/2014/06/a-sleek-new-hearing-aid-that-solves-a-nagging-problem/.

38. "Introducing Soundhawk," YouTube video, accessed September 1, 2019, https://youtu.be/P3ChmlkmKXI.

39. David Pierce, "Inside the Downfall of Doppler Labs," *Wired*, November 1, 2017, www.wired.com/story/inside-the-downfall-of-doppler-labs/.

40. "Here One," home page, accessed October 16, 2017, hereplus.me.

41. CNET, "Here One 'Smart' Wireless Earphones Aren't AirPod Killers, but They're better in Some Ways," posted February 21, 2017, YouTube video, www.youtube.com/watch?v=XoJWDcSzwsI.

42. "Product Details," Here One, accessed October 26, 2017, https://hereplus.me/pages/product-details.

43. Hagood, *Hush*.

44. Ryan Waniata, "Here One Review," *Digital Trends*, January 1, 2017, www.digitaltrends.com/headphone-reviews/Doppler-labs-here-one-review.

45. Waniata, "Here One Review."

46. David Pierce, "Bragi Dash Puts a New Kind of Computer in Your Ears," *Wired*, January 11, 2016, www.wired.com/2016/01/bragi-dash/.

47. Pierce, "New Computer in Your Ears."

48. To see footage of the Bragi Dash in use, see The Verge, "Bragi Dash Wireless Earbuds Review," posted March 18, 2016, YouTube video, www.youtube.com/watch?v=BTLhQ11snU8.

49. Will Oremus, "These Aren't Wireless Headphones," *Slate*, September 8, 2016, www.slate.com/articles/technology/future_tense/2016/09/apple_s_airpods_aren_t_just_wireless_earbuds_they_re_the_future_of_computing.html.

50. "Design for Everyone," Apple Developer, 2017, video, https://developer.apple.com/videos/play/wwdc2017/806/.

51. Heidi Rae Cooley, *Finding Augusta: Habits of Mobility and Governance in the Digital Era* (Hanover, NH: Dartmouth College Press, 2014).

52. David Parisi, "Game Interfaces as Disabling Infrastructures," *Analog Game Studies* 4, no. 3 (2017), http://analoggamestudies.org/2017/05/compatibility-test-videogames-as-disabling-infrastructures/.

53. Parisi, "Game Interfaces Disabling Infrastructures."

54. Katie Ellis and Mike Kent, *Disability and New Media* (New York: Routledge, 2011), 36.

55. Alper, *Giving Voice*.

56. Steven Aquino, "AirPods to Get Live Listen Feature in iOS 12," *Tech Crunch*, June 5, 2018, https://techcrunch.com/2018/06/05/airpods-to-get-live-listen-feature-in-ios-12/.

57. Steven Aquino, "At Apple's WWDC 2018, Accessibility Pervades All," *Tech Crunch*, June 7, 2018, https://techcrunch.com/2018/06/07/at-apples-wwdc-2018-accessibility-pervades-all/.

58. For more on trade rituals from a media industries perspective, see John Thornton Caldwell, *Production Culture: Industrial Reflexivity and Critical Practice in Film and Television* (Durham, NC: Duke University Press, 2006).

59. David T. Mitchell and Sharon L. Snyder, *The Biopolitics of Disability: Neoliberalism, Ablenationalism, and Peripheral Embodiments* (Ann Arbor: University of Michigan Press, 2015), 5.

60. Elizabeth Ellcessor, *Restricted Access: Media, Disability, and the Politics of Participation* (New York: New York University Press, 2006), 7.

61. Ellcessor, *Restricted Access*, 8.

62. Mitchell and Snyder, *Biopolitics of Disability*.

CHAPTER 3: SPORTS, OR: MONITORING PHYSICAL ACTIVITY ON AND OFF THE FIELD

1. Michael Lewis, *Moneyball: The Art of Winning an Unfair Game* (New York: W.W. Norton and Company, 2004), 83 (e-book version).

2. Lewis, *Moneyball.*

3. Nicholas Negroponte, *Being Digital* (New York: Vintage, 1996).

4. Neil Smith, "Homeless/Global: Scaling Places," in *Mapping the Futures: Local Cultures, Global Change*, ed. John Bird et al. (New York: Routledge, 1993), 87–119. For Smith, using "scale" in such a way shows us how objects transform depending on our vantage point of thinking about them.

5. Sarah M. Brown and Katie M. Brown, "Should Your Wearables Be Shareable? The Ethics of Wearable Technology in Collegiate Athletics," *Marquette Sports Law Review* 32, no. 1 (2021): 97–116.

6. Rochelle Nicholls et al., "Baseball: Accuracy of Qualitative Analysis for Assessment of Skilled Baseball Pitching Technique," *Sports Biomechanics* 2, no. 2 (2003): 213–226.

7. Deborah Lupton, Sarah Pink, and Christine Heyes LaBond, "Digital Traces in Context: Personal Data Contexts, Data Sense, and Self-Tracking Cycling," *International Journal of Communication* 12 (2018): 647–665.

8. Mark McClusky, "The Nike Experiment: How the Shoe Giant Unleashed the Power of Personal Metrics," *Wired*, June 22, 2009, www.wired.com/2009/06/lbnp-nike/. For more information on the Nike+iPod kit in research on running, see Philippa Trevorrow, "Technology Running the World: The Nike+iPod Kit and Levels of Physical Activity," *Society and Leisure* 35, no. 1 (2012): 131–154.

9. Walter S. Mossberg and Katherine Boehret, "On the Run with the iPod/Nike Fitness Device," *Wall Street Journal*, July 16, 2006, www.wsj.com/articles/SB115326608907010478.

10. Apple, "Nike and Apple Team Up to Launch Nike+iPod," Newsroom, May 23, 2006, www.apple.com/newsroom/2006/05/23Nike-and-Apple-Team-Up-to-Launch-Nike-iPod/.

11. "Nike iPod Tune Your Run OK Go," posted December 1, 2006, YouTube video, www.youtube.com/watch?v=HEs8NIRRyYc.

12. McClusky, "Nike Experiment."

13. Further, the development of standards and baselines out of that data echoes earlier modes of establishing baselines through bodily observation, as recounted in David Armstrong, "The Rise of Surveillance Medicine," *Sociology of Health & Illness* 17, no. 3 (1995): 393–404.

14. Lisa Gitelman, *"Raw Data" Is an Oxymoron* (Cambridge, MA: MIT Press, 2013).

15. BeetTV, "TechCrunch 50 Runner-Up Fitbit Tracks Activities," Posted September 12, 2008, YouTube video, www.youtube.com/watch?v=oW3JRr9PIHg.

16. John Biggs, "TC50: FitBit, a Fitness Gadget That Makes Us Want to Exercise," *TechCrunch*, September 9, 2008, https://techcrunch.com/2008/09/09/tc50-fitbit-fitness-gadget-the-makes-us-want-to-exercise.

17. "Tips for Getting Your Steps In," Centers for Disease Control and Prevention, May 9, 2016, www.cdc.gov/features/getting-your-steps-in/index.html.

18. In the words of Charles Peirce, this is more of an iconic rather than an indexical representation. See Charles Peirce, *Collected Writings*, ed. Charles Hartshorne,

Paul Weiss, and Arthur Banks (Cambridge, MA: Harvard University Press, 1965). It is also important to note here that the relationship between technology and health is different from that explored in chapter 1. Some organizations, such as the US Department of Veterans Affairs, have suggested using wearable trackers to measure a person's activity after receiving a new prosthetic in order to give doctors a way to more precisely track patients while also compelling them to work on physical therapy. See Jonah Comstock, "VA to Reimburse for Certain Clinical Activity Tracks," *Mobi Health News*, August 28, 2014, www.mobihealthnews.com/36158/va-to-reimburse-for-certain-clinical-activity-trackers.

19. Fitbit's mode of "sharing" resonates with the way Nicholas A. John describes the "Age of Sharing" as defined by a communicative definition of *share*, in which the act of sharing is bound up in messages and representations and less in the distributive idea of sharing something material by splitting it into pieces. See Nicholas A. John, *The Age of Sharing* (Malden, MA: Polity, 2016).

20. And it may be a key reason these devices interact so much with everyday life, a domain that has been argued to be interpersonally constructed. See Peter Berger and Thomas Luckmann, *The Social Construction of Reality: A Treatise on the Sociology of Knowledge* (New York: Anchor Books, 1967).

21. This offers something of an update on Karl Marx's suggestion that physical forces enter into the process of values and can become transactional. Marx's emphasis was on machines of production and how they shift values of labor and society, but Fitbit offers a way to see how different sorts of machines and devices transform physical capital—walking and running—into social capital—leader boards, Fitbit friends, and sending "cheers" over the application. See Karl Marx, *Capital*, vol. 1, *A Critique of Political Economy* (New York: Penguin, 1992).

22. See also Pierre Bourdieu, *Distinction: A Social Critique of the Judgement of Taste* (New York: Routledge, 1984).

23. Jennifer R. Whitson, "Gaming the Quantified Self," *Surveillance and Society* 11, nos. 1–2 (2013): 164.

24. Whitson, "Gaming the Quantified Self," 166.

25. Whitson, "Gaming the Quantified Self," 169.

26. Ian Bogost, "Why Gamification Is Bullshit," in *The Gamified World: Approaches, Issues, Applications*, ed. Steffen P. Walz and Sebastian Deterding (Cambridge, MA: MIT Press, 2014), 68.

27. Gilles Deleuze, "Postscript on the Societies of Control," *October* 59 (1992): 5.

28. As Martin French and Gavin Smith have suggested, "health context[s]" like running act as "test bed[s] for novel data-extraction technologies and practices," which can "legitimatize their value and stature in the public imaginary." Martin French and Gavin Smith, "'Health' Surveillance: New Modes of Monitoring Bodies, Populations, and Polities," *Critical Public Health* 23, no. 4 (2013): 384. See further Martin French and Gavin J. D. Smith, "Surveillance and Embodiment: Dispositifs of Capture," *Body & Society* 22, no. 2 (2016): 3–27.

29. Jeremy Hsu, "The Strava Heat Map and the End of Secrets," *Wired*, January 29, 2018, www.wired.com/story/strava-heat-map-military-bases-fitness-trackers-privacy/.

30. Drew Robb, "Building the Global Heatmap," Medium, November 1, 2017, https://medium.com/strava-engineering/the-global-heatmap-now-6x-hotter-23fc01d301de.

31. Alex Hern, "Fitness Tracking App Strava Gives Away Location of Secret US Army Bases," *Guardian*, January 28, 2018, www.theguardian.com/world/2018/jan/28/fitness-tracking-app-gives-away-location-of-secret-us-army-bases.

32. For instance, Sun Joo (Grace) Ahn, Kyle Johnsen, and Catherine Ball, "Points-Based Reward Systems in Gamification Impact Children's Physical Activity Strategies and Psychological Needs," *Health Education & Behavior* 46, no. 3 (2019): 417–425; and Rungting Tu, Peishan Hsieh, and Wenting Feng, "Walking for Fun or for 'Likes'? The Impacts of Different Gamification Orientations of Fitness Apps on Consumers' Physical Activities," *Sport Management Review* 22, no. 5 (2019): 682–693.

33. This generally speaks to a transformation in Foucault's work from a focus on the technique of power he described in *Discipline and Punish* to a focus on the constitution of human subjects within different power matrices of social, historical, economic, and moral frameworks, such as that traced across his three-part *History of Sexuality*. See Michel Foucault, *Discipline and Punish: The Birth of the Prison* (New York: Vintage, 1995); and Michel Foucault, *History of Sexuality*, vol 3, *The Care of the Self* (New York: Vintage, 1988).

34. Mark G. E. Kelly, "Foucault, Subjectivity, and Technologies of the Self," in *A Companion to Foucault*, ed. Christopher Falzon, Timothy O'Leary, and Jana Sawicki (Hobocken, NJ: Blackwell, 2013), 510–525.

35. See, for instance, Katleen Gabriels and Mark Coeckelbergh, "'Technologies of the Self and Other': How Self-Tracking Technologies Also Shape the Other," *Journal of Information, Communication and Ethics in Society* 17, no. 2 (2019): 119–127; Dušan Ristić and Dušan Marinkovič, "Lifelogging: Digital Technologies of the Self as Practices of Contemporary biopolitics," *Siologija* 4 (2019): 535–549; and Meghan McMahon, "Grappling with the Datafied Self: College Students and Wearable Fitness Trackers," *Columbia University Journal of Politics and Society* 31, no. 1 (2020): 56–88.

36. Elizabeth Chuck, "Oral Roberts University to Track Students' Fitness through Fitbits," *NBC News*, February 3, 2016, www.nbcnews.com/feature/college-game-plan/oral-roberts-university-track-students-fitness-through-fitbits-n507661.

37. Associated Press, "ORU Students Log 3 Billion Steps on Fitbit Devices," March 20, 2017, www.apnews.com/4574909b130d475089f6970f816d9f65.

38. Deleuze, "Postscript on societies of control," 3–7.

39. Deleuze, "Postscript on societies of control," 3–7.

40. "New Student Orientation Schedule," Oral Roberts University, 2018, www.oru.edu/oru-experience/first-year-experience/orientation/schedule.php.

41. Michel Foucault, *Power/Knowledge: Selected Interviews and Other Writings, 1972–1977* (New York: Pantheon, 1980).

42. Jeff Stone, "Not All Oral Roberts Students Need to Wear Fitbits, and They're Not Tracked through Campus," *International Business Times*, February 3, 2016, www.ibtimes.com/not-all-oral-roberts-students-need-wear-fitbits-theyre-not-tracked-through-campus-2291808, emphasis added.

43. Claire Landsbaum, "Why a Christian University's Freshman Fitbit Requirement Is a Bad Idea," *New York Magazine*, February 4, 2016, http://nymag.com/scienceofus/2016/02/christian-school-requires-fitbits-for-freshmen.html.

44. This is not dissimilar from many ways datafication has been used to track students in educational settings. For more, see Ben Williamson, *Big Data in Education: The Digital Future of Learning, Policy, and Practice* (Thousand Oaks, CA: Sage Publications, 2017); and Deborah Lupton and Ben Williamson, "The Datafied Child: The Dataveillance of Children and Implications for Their Rights," *New Media & Society* 19, no. 5 (2017): 780–794.

45. Sara Gregory, "The New Heroes of High School Gym Class: Fitness Trackers," *Roanoke Times*, February 21, 2018, www.roanoke.com/news/education/the-new-heroes-of-high-school-gym-fitness-trackers/article_0ba3f366-384c-5e40-a578-09187e575ac7.html.

46. Gregory, "Heroes of High School Gym Class."

47. For an overview of the technical system and how it works in physical education courses, see: Polar, "Making Physical Education Measurably More Fun," accessed June 20, 2018, www.polar.com/en/business/education/

48. Gregory, "Heroes of High School Gym Class."

49. Polar, "Privacy Notice," accessed June 20, 2018, www.polar.com/us-en/legal/privacy-notice#toc25.

50. Victoria Song, "Only Athletes Should Give a Whoop about Whoop," *Verge*, March 2, 2022, www.theverge.com/22957195/whoop-review-fitness-tracker-wearables.

51. Marc Tracy, "Technology Used to Track Players' Steps Now Charts Their Sleep, Too," *New York Times*, September 22, 2017, www.nytimes.com/2017/09/22/sports/ncaafootball/clemson-alabama-wearable-technology.html.

52. Ideo, "A Game-Changing Approach to Sleep for Athletes," accessed January 10, 2023, www.ideo.com/works/a-game-changing-approach-to-sleep-for-athletes

53. Marc Tracy, "With Wearable Tech Deals, New Player Data Is Up for Grabs," *New York Times*, September 11, 2016, www.nytimes.com/2016/09/11/sports/ncaafootball/wearable-technology-nike-privacy-college-football.html.

54. Tracy, "Technology Used to Track Players."

55. The idea of "culture of surveillance" comes from David Lyon, who has argued that surveillance practices have become so pervasive and endemic to everyday life that individuals accept them largely unquestioningly and often see them as useful, even if those practices are being used to police, discipline, or control them. See David Lyon, *The Culture of Surveillance: Watching as a Way of Life* (Malden, MA: Polity, 2018).

56. The controversy over NIL (name, image, and likeness) in NCAA athletes demonstrates this, as well. Student-athletes successfully lobbied for the ability to generate revenue off their own individual name and brand, both in relation to their institution and as individual athletes. Jonathan E. Howe, Wayne L. Black, and Willis A. Jones. "Exercising Power: A Critical Examination of National Collegiate Athletic Association Discourse Related to Name, Image, and Likeness," *Journal of Sport Management* 37, no. 5 (2023): 1–12.

57. Barry Petchesky, "Auburn Has a Private Security Firm Enforcing Players' Nightly Curfews," *Deadspin* (blog), November 8, 2012, https://deadspin.com/auburn-has-a-private-security-firm-enforcing-players-ni-5958852.

58. Jake New, "Class Checkers," *Inside Higher Ed*, June 23, 2015, www.insidehighered.com/news/2015/06/24/attendance-monitoring-programs-common-college-athletics.

59. Lauren McCoy, "You Have the Right to Tweet, but It Will Be Used Against You: Balancing Monitoring and Privacy for Student-Athletes," *Journal of SPORT* 3, no. 2 (2014): 221–245.

60. At a time when increased attention is being paid to "the disproportionately racialized nature of [collegiate athletics'] exploitative dynamics," there are also undoubtedly racial politics at play here. Nathan Kalman-Lamb, Derek Silva, and Johanna Mellis, "'I Signed My Life to Rich White Guys': Athletes on the Racial Dynamics of College Sports," *Guardian*, March 17, 2021, www.theguardian.com/sport/2021/mar/17/college-sports-racial-dynamics.

61. Tracy, "With Wearable Tech Deals."

62. David Hale, "FSU Ride GPS Technology to Title," *ESPN*, June 22, 2014, www.espn.com/college-football/story/_/id/11121315/florida-state-seminoles-coach-jimbo-fisher-use-gps-technology-win-national-championship.

63. Hale, "FSU Ride GPE Technology."

64. Rainer Sabin, "Inside the Technology Giving Alabama a Competitive Edge," Alabama Football, July 2, 2017, www.al.com/alabamafootball/2017/07/inside_the_technology_giving_a.html.

65. Alabama Crimson Tide on AL.com, "Alabama Football Uses Top of the Line GPS Technology to Track Player Performance," posted July 2, 2017, YouTube video, www.youtube.com/watch?v=DT7Yk7ED1QM.

66. Catapult, "Catapult Teams Dominate College Football Standings," *PR Newswire*, December 21, 2021, www.prnewswire.com/news-releases/catapult-teams-dominate-college-football-standings-301449189.html.

67. 3"Athletics Partners with WHOOP to Support Student-Athlete Holistic Wellness," Penn State Athletics, August 16, 2022, https://gopsusports.com/news/2022/8/16/general-athletics-partners-with-whoop.aspx.

68. Joe Lemire, "Don't Sleep on the 'U': Miami Football Players Used Own It and Whoop to Get Their Rest and Woke Up Their 2021 Season," *Sports Business Journal*, April 29, 2022, www.sportsbusinessjournal.com/Daily/Issues/2022/04/29/Technology/dont-sleep-on-the-u-miami-football-players-set-example-by-using-own-it-and-whoop-to-get-their-rest-and-wake-up-their-2021-season.aspx.

69. Jimmy Sanderson et al., "'I Was Able to Still Do My Job on the Field and Keep Playing': An Investigation of Female and Male Athletes' Experiences with (Not) Reporting Concussions," *Communication and Sport* 5, no. 3 (2017): 267–287.

70. Alicia Jessop and Thomas A. Baker II, "Big Data Bust: Evaluating the Risks of Tracking NCAA Athletes' Biometric Data," *Texas Review of Entertainment and Sports Law* 20, no. 1 (Fall 2019): 81–112.

71. Daily Collegian, The, "WHOOP There It Is: James Franklin Explains Penn State Football's Wearable Sleep Monitor," posted September 13, 2017, YouTube video, www.youtube.com/watch?v=FmcScmytHg8.

72. For an overview of surveillance ethics as they are applied to "care," see Eric Stoddard, "A Surveillance of Care: Evaluating Surveillance Ethically," in *Routledge Handbook of Surveillance Studies*, ed. Kirstie Ball, Kevin Haggerty, and David Lyon (New York: Routledge), 369–376. For health communication contexts, see Claudia Lang, "Inspecting Mental Health: Depression, Surveillance, and Care in Kerala, South India," *Culture, Medicine and Psychiatry* 43 (2019): 596–612; and Shin Y. Kim, Nicholas P. Deputy, and Cheryl L. Robbins, "Diabetes during Pregnancy: Surveillance, Preconception Care, and Postpartum Care," *Journal of Women's Health* 27, no. 5 (2018): 536–541.

73. For an example, see Krystal K. Beamon, "'Used Goods': Former African American College Student-Athletes' Perception of Exploitation by Division I Universities," *Journal of Negro Education* 77, no. 4 (2008): 352–364.

74. Dorothy Nelkin and Lori Andrews, "DNA Identification and Surveillance Creep," *Sociology of Health and Illness* 21, no. 5 (1999): 689–706.

75. Brittany Ghiroli and Rob Biertempfel, "Baseball's New Frontier: How Wearable Technology Is Reshaping the Game," *Athletic*, March 21, 2019, https://theathletic.com/879248/2019/03/21/baseballs-new-frontier-how-wearable-technology-is-reshaping-the-game/.

76. John A. Balletta, "Measuring Baseball's Heartbeat: The Hidden Harms of Wearable Technology to Professional Ballplayers," *Duke Law and Technology Review* 18, no. 1 (2019): 268–292.

77. Louis Althusser, *On the Reproduction of Capitalism* (New York: Verso, 2016).

CHAPTER 4: LABOR, OR: WORKPLACE SURVEILLANCE DOWN TO THE MILLISECOND

1. Ceylan Yeginsu, "If Workers Slack Off, the Wristband Will Know (and Amazon Has a Patent for It)," *New York Times*, February 1, 2018, www.nytimes.com/2018/02/01/technology/amazon-wristband-tracking-privacy.html.

2. Jonathan Evan Cohn, "Ultrasonic Bracelet and Receiver for Detecting Position in 2D Plane," Google Patents, accessed May 20, 2018, https://patents.google.com/patent/US20170278051A1/en, p. 1

3. Yeginsu, "The Wristband Will Know."

4. Heather Kelly, "Amazon's Idea for Employee-Tracking Wearables Raises Concerns," *CNN*, February 2, 2018, https://money.cnn.com/2018/02/02/technology/amazon-employee-tracker/index.html.

5. Thuy Ong, "Amazon Patents Wristbands That Track Warehouse Employees' Hands in Real Time," *Verge*, February 1, 2018, www.theverge.com/2018/2/1/16958918/amazon-patents-trackable-wristband-warehouse-employees.

6. For a description of these warehouse robots, see Matt Simon, "Inside the Amazon Warehouse Where Humans and Machines Become One," *Wired*, June 5, 2019, www.wired.com/story/amazon-warehouse-robots/.

7. Sarah Butler, "Amazon Accused of Treating UK Warehouse Staff Like Robots," *Guardian*, May 31, 2018, www.theguardian.com/business/2018/may/31/amazon-accused-of-treating-uk-warehouse-staff-like-robots; Emily Guendelsberger, "I Worked at an Amazon Fulfillment Center; They Treat Workers Like Robots," *Time*, July 18, 2019, https://time.com/5629233/amazon-warehouse-employee-treatment-robots/; and Josh Dzieza, "'Beat the Machine': Amazon Warehouse Workers Strike to Protest Inhumane Conditions," *Verge*, July 16, 2019, www.theverge.com/2019/7/16/20696154/amazon-prime-day-2019-strike-warehouse-workers-inhumane-conditions-the-rate-productivity.

8. For more about the "we are not robots!" declarations, see Alessandro Delfanti, "Machinic Dispossession and Augmented Despotism: Digital Work in an Amazon Warehouse," *New Media & Society* 23, no. 1 (2019): 39–55.

9. For information on e-mail and web browsing surveillance, see Esther Milne, *Email and the Everyday: Stories of Disclosure, Trust, and Digital Labor* (Cambridge, MA: MIT Press, 2021).

10. Melissa Gregg, *Counterproductive: Time Management in the Knowledge Economy* (Durham, NC: Duke University Press, 2018); and Melissa Gregg, *Work's Intimacy* (Malden: Polity, 2011).

11. For reportage on the Lab, see Stewart Brand, *The Media Lab: Inventing the future at MIT* (New York: Viking Books, 1987).

12. Nicholas Negroponte, *Being Digital* (London: Hodder and Stoughton, 1995), 4.

13. Morgan G. Ames, *The Charisma Machine: The Life, Death, and Legacy of One Laptop per Child* (Cambridge, MA: MIT Press, 2019).

14. Katherine Noyes, "Humanyze's 'People Analytics' Wants to Transform Your Workplace," *Computer World*, November 20, 2015, www.computerworld.com/article/3006631/startup-humanyzes-people-analytics-wants-to-transform-your-workplace.html.

15. Jordan Frith, *A Billion Little Pieces: RFID and Infrastructures of Identification* (Cambridge, MA: MIT Press, 2019), 6.

16. Mark Andrejevic and Mark Burdon, "Defining the Sensor Society," *Television and New Media* 16, no. 1 (2015): 19–36.

17. Ben Waber, *People Analytics: How Social Sensing Technology will Transform Business and What it Tells us About the Future of Work*. Upper Saddle River, NJ: FT Press, 2013.

18. Ben Waber, *People Analytics*, e-book, O'Reilly learning platform, accessed September 22, 2019, https://learning.oreilly.com/library/view/people-analytics-how/9780133158342/toc.html.

19. Ellen Sheng, "Employee Privacy in the US Is at Stake as Corporate Surveillance Technology Monitors Workers' Every Move," *CNBC*, April 15, 2019, www.cnbc.com/2019/04/15/employee-privacy-is-at-stake-as-surveillance-tech-monitors-workers.html.

20. MIT Technology Review Insights, "Technology for Workplaces That Work: Humanyze's Ben Waber," *MIT Technology Review*, January 24, 2019, www.technologyreview.com/2019/01/24/137732/technology-for-workplaces-that-work-humanyzes-ben-waber/.

21. Humanyze, "Company," accessed October 1, 2019, www.humanyze.com/about/ (emphasis in original).

22. Ben Waber, "Using Analytics to Measure Interactions in the Workplace," re:Work with Google, posted November 10, 2014, YouTube video, www.youtube.com/watch?v=XojhyhoRI7I.

23. Ron Miller, "New Firm Combines Wearables and Data to Improve Decision Making," *Tech Crunch*, February 24, 2015, https://techcrunch.com/2015/02/24/new-firm-combines-wearables-and-data-to-improve-decision-making/.

24. Robin Young, "Beyond Counting Footsteps: Wearable Tech That Measures How You Work," WBUR, November 9, 2015, www.wbur.org/hereandnow/2015/11/09/humanyze-measuring-work-habits.

25. For some discussion on the politics of aggregate data as an alibi for data collection, see, for instance, Mark Andrejevic, "Surveillance in the Digital Enclosure," *Communication Review* 10, no. 4 (2007): 295–317; and Zeynep Tufekci, "Engineering the Public: Big Data, Surveillance and Computational Publics," *First Monday* 19, no. 7 (2014), https://journals.uic.edu/ojs/index.php/fm/article/view/4901.

26. Chris Weller, "Employees at a Dozen Fortune 500 Companies Wear Digital Badges That Watch and Listen to Their Every Move," *Business Insider*, October 21, 2016, www.businessinsider.com/humanyze-badges-watch-and-listen-employees-2016-10.

27. Natasha Lomas, "Researchers Spotlight the Lie of 'Anonymous' Data," *Tech Crunch*, July 24, 2019, https://techcrunch.com/2019/07/24/researchers-spotlight-the-lie-of-anonymous-data/; Yves-Alexandre de Montjoyve et al., "Unique in the Shopping Mall: On the Reidentifiability of Credit Card Metadata," *Science* 347, no. 6221 (2015): 536–539; and Jonathan Mayer, Patrick Mutchler, and John C. Mitchell, "Evaluating the Privacy Properties of Telephone Metadata," *Proceedings of the National Academy of Sciences of the United States of America* 113, no. 20 (2016): 5536–5541.

28. Pentland has also developed his theory of "social physics," which aims to in part use large-scale data tracking to assess human interactions. Alex Pentland, *Social Physics: How Social Networks Can Make Us Smarter* (New York: Penguin Books, 2015).

29. There are affinities here with Luke Stark's exploration of the "scalable subject" as a means of understanding how companies use aggregate metadata to make inferences about generalizable human behavior. Luke Stark, "Algorithmic Psychometrics and the Scalable Subject," *Social Studies of Science* 48, no. 2 (2018): 204–231.

30. For discussion of how Chaplin's work represented and critiqued these systems, see Tom Gunning, "Chaplin and the Body of Modernity," *Early Popular Visual Culture* 8, no. 3 (2010): 237–245.

31. Nikil Saval, *Cubed: A Secret History of the Workplace* (New York: Doubleday, 2014).

32. Phoebe V. Moore, *The Quantified Self in Precarity: Work, Technology, and What Counts* (New York: Routledge, 2018), 52.

33. Moore, *Quantified Self in Precarity*, 63.

34. Gregg, *Counterproductive.*

35. For more discussion of Taylorist principles, including the stopwatch experiments, and their connection to things like the quantified self, see Christopher O'Neill, "Taylorism, the European Science of Work, and the Quantified Self at Work," *Science, Technology, and Human Values* 42, no. 4 (2017): 600–621.

36. E. P. Thompson, "Time, Work-Discipline, and Industrial Capitalism," *Past & Present* 38 (1967): 61.

37. Thompson, "Time, Work-Discipline," 90.

38. Esther Leslie, "This Other Atmosphere: Against Human Resources, Emoji and Devices," *Journal of Visual Culture* vol. 18, no. 1 (2019): 5.

39. Leslie, "This Other Atmosphere," 12.

40. Rachel Hall, Torin Monahan, and Joshua Reeves have offered a similar point in their overview of the tensions between "performance monitoring" and "monitored performance." See Rachel Hall, Torin Monahan, and Joshua Reeves, "Editorial: Surveillance and Performance," *Surveillance & Society* 14, no. 2 (2016): 154–167.

41. Alfred D. Chandler Jr., *The Visible Hand: The Managerial Revolution in American Business* (Cambridge, MA: Harvard University Press, 1993). See also John Gilliom and Torin Monahan, *SuperVision: An Introduction to the Surveillance Society* (Chicago: University of Chicago Press, 2012).

42. This sense of the geneaological I take largely from Michel Foucault, who likewise was interested in examining historical transformations of knowledge. See, for an example of this, *The Birth of the Clinic: An Archaeology of Medical Perception* (New York: Routledge, 1989).

43. Kirstie Ball, "Workplace Surveillance: An Overview," *Labor History* 51, no. 1 (2010): 89.

44. Michel Foucault, *The Order of Things: An Archaeology of Human Sciences* (New York: Vintage Books, 1994).

45. Waber, *People Analytics*, e-book.

46. Ifeoma Ajunwa, Kate Crawford, and Jason Schultz, "Limitless Worker Surveillance," *California Law Review* 105 (2017): 770.

47. Greg Lindsay, "We Spent Two Weeks Wearing Employee Trackers: Here's What We Learned," *Fast Company*, September 9, 2015, www.fastcompany.com/3051324/we-spent-two-weeks-wearing-employee-trackers-heres-what-we-learned.

48. Lindsay, "Two Weeks Wearing Trackers."

49. Lindsay, "Two Weeks Wearing Trackers."

50. Lindsay, "Two Weeks Wearing Trackers."

51. Lindsay, "Two Weeks Wearing Trackers."

52. The idea of "information-discipline" comes from Gilmore's work on the early iterations of smartwatches. There, it was used to describe the need to manage information flows on smartwatches; here, it is connected more to a panoptic mode of power in which one's data becomes visible and needs to be modulated. See James N. Gilmore, "From Ticks and Tocks to Budges and Nudges: The Smartwatch and the Haptics of Informatic Culture," *Television & New Media* 18, no. 3 (2017): 189–202.

53. Ben Waber, "Decoding Workforce Productivity," World Economic Forum, posted August 8, 2016, YouTube video, www.youtube.com/watch?v=i-F7Cd_W4Uc.

54. Young, "Beyond Counting Footsteps."

55. Miller, "New Firm Combines Wearables."

56. Rosalind Picard, *Affective Computing* (Cambridge, MA: MIT Press, 2000); and Blake Hallinan, "Civilizing Infrastructure," *Cultural Studies* 35, nos. 4–5 (2021): 707–727.

57. MIT Technology Review Insights, "Technology for Workplaces."

58. MIT Technology Review Insights, "Technology for Workplace.".

59. David Beer, *The Data Gaze: Capitalism, Power, and Perception* (Thousand Oaks, CA: Sage Publications, 2018).

60. Young, "Beyond Counting Footsteps."

61. Noyes, "Humanyze's 'People Analytics.'"

62. Rosalie Chan, "The High-Tech Office of the Future Will Spy on You," *Week*, June 6, 2018, https://theweek.com/articles/760582/hightech-office-future-spy.

63. Winifred Poster, "Emotion Detectors, Answering Machines and E-unions: Multisurveillance in the Global Interactive Services Industry," *American Behavioral Scientist* 55, no. 7 (2011): 868–901; Ruha Benjamin, *Race after Technology: Abolitionist Tools for the New Jim Code* (Malden, MA: Polity, 2019); and Virginia Eubanks, *Automating Inequality: How High-Tech Tools Profile, Police, and Punish the Poor* (New York: St. Martin's Press, 2018).

64. For more examples of this kind of reporting, see Thomas Heath, "Employee ID Badge Monitors and Listens to You at Work—Except in the Bathroom," *Washington Post*, September 7, 2016, www.washingtonpost.com/news/business/wp/2016/09/07/this-employee-badge-knows-not-only-where-you-are-but-whether-you-are-talking-to-your-co-workers/; and Scott Snyder and Alex Castrounis, "How to Turn 'Data Exhaust' into a Competitive Edge," *Knowledge at Wharton*, March 1, 2018, https://knowledge.wharton.upenn.edu/article/turn-iot-data-exhaust-next-competitive-advantage/.

65. Miller, "New Firm Combines Wearables."

66. Gregg, *Counterproductive*, 46.

67. Waber, *People Analytics*, e-book.

68. Raymond Williams, *Keywords: A Vocabulary of Culture and Society*, rev. ed. (New York: Oxford University Press, 1983).

69. Mark Mann, "How Companies Are Using Technology to Make Workers 'Happy' in Their Crappy Jobs," *Vice*, September 5, 2016, www.vice.com/en_us/article/aekn4a/big-data-social-physics-humanyze-tenacity-employment.

70. Ted Striphas, "Algorithmic Culture," *European Journal of Cultural Studies* 18, nos. 4–5 (2015): 397. Striphas also traces this connection to Friedrich Kittler, "Thinking Colours and/or Machines," *Theory, Culture, & Society* 23, nos. 7–8 (2006): 39–50.

71. C. P. Snow once summed this up as "the two cultures" to describe how scientific and humanist modes of research pit themselves against each other and make claims to authority. C. P. Snow, *The Two Cultures* (Cambridge: Cambridge University Press, 2014).

72. Stuart Hall wrote about this at length in his lectures on Antonio Gramsci and the theory of hegemony. For more, see Stuart Hall, *Cultural Studies 1983: A Theoretical History* (Durham, NC: Duke University Press, 2018).

73. This idea serves as the basis of Raymond Williams's work in *Culture and Society: 1780–1950* (New York: Anchor Books, 1960).

74. Here, I mean to think about programming as a logic of control that has been discussed in relation to computer protocol and humans' learned relationships to computer systems. See Alexander Galloway, *Protocol: How Control Exists after Decentralization* (Cambridge, MA: MIT Press, 2004).

75. Thompson, "Time, Work-Discpline."

76. For more information, see, for instance, Joseph Fuller and William Kerr, "The Great Resignation Didn't Start with the Pandemic," *Harvard Business Review*, March 23, 2022, https://hbr.org/2022/03/the-great-resignation-didnt-start-with-the-pandemic.

77. Humanyze, "Case Study: Technology Company Measures the Impacts of Remote Work to Drive Organizational Health," 2020, http://humanyze.wpengine.com/wp-content/uploads/2020/12/Technology-Company-Measures-Impacts-of-Remote-Work-Case-Study.pdf.

78. Humanyze, "Case Study."

79. Ben Waber, "Don't Let COVID-19 Compromise Your Organizational Health," Humanyze, March 10, 2021, https://humanyze.com/dont-let-covid-19-compromise-your-organizational-health/.

80. Ben Waber, "How Companies Can Maintain Organizational Health amid COVID-19," *TLNT*, May 27, 2020, www.tlnt.com/articles/how-companies-can-maintain-organizational-health-amid-covid-19.

81. Ben Waber, "Work-Life Balance in the Time of Coronavirus," *Humanyze* (blog), accessed November 10, 2022, https://humanyze.com/blog-work-life-balance-during-coronavirus/.

82. Rahina Andre, "Humanyze Announces New Workplace Strategy Solution to Help Companies Inform and Improve Workplace Decisions with Science-Backed

Insights," Humanyze, August 17, 2021, https://humanyze.com/humanyze-announces-new-workplace-strategy-solution-to-help-companies-inform-and-improve-workplace-decisions-with-science-backed-insights/.

CHAPTER 5: LAW ENFORCEMENT, OR: THE OPACITY OF BODY-WORN CAMERAS IN UPSTATE SOUTH CAROLINA

1. Michael S. Schmidt and Matt Apuzzo, "South Carolina Officer Is Charged with Murder of Walter Scott," *New York Times*, April 8, 2015, www.nytimes.com/2015/04/08/us/south-carolina-officer-is-charged-with-murder-in-black-mans-death.html.

2. Steve Mann, "Sousveillance: Inventing and Using Wearable Computing Devices for Data Collection in Surveillance Environments," *Surveillance & Society* 1, no. 3 (2003), doi.org/10.24908/ss.v1i33344; and Steve Mann, "New Media and the Power Politics of Sousveillance in a Surveillance-Dominated World," *Surveillance & Society* 11, nos. 1–2 (2013): 18–34.

3. The scholarship on Black Lives Matter is vast, but several sources that provide context in relationship to the discussions of this chapter are Christopher J. Lebrun, *The Making of Black Lives Matter: A Brief History of an Idea* (New York: Oxford University Press, 2017); Alondra Nelson, "The Longue Duree of Black Lives Matter," *American Journal of Public Health* 106, no. 10 (2016): 1734–1737; Jennifer Chernega, "Black Lives Matter: Racialized Policing in the United States," *Comparative American Studies* 14, nos. 3–4 (2016): 234–245; and Jonathan Havercroft, "Soul-Blindness, Police Orders, and Black Lives Matter," *Political Theory* 44, no. 6 (2016): 739–763.

4. Alan Blinder, "Michael Slager, Officer in Walter Scott Shooting, Gets 20-Year Sentence," *New York Times*, December 7, 2017, www.nytimes.com/2017/12/07/us/michael-slager-sentence-walter-scott.html.

5. Wesley Bruer, "Obama Warns Cop Body Cameras Are No 'Panacea,'" *CNN*, March 3, 2015, www.cnn.com/2015/03/02/politics/obama-police-body-camera-report/index.html.

6. C .J. Reynolds, "Mischievous Infrastructure: Tactical Secrecy through Infrastructural Friction in Police Video Systems," *Cultural Studies* 35, nos. 4–5 (2021): 996–1019.

7. Apart from the legal recourse these videos provided for the police who killed these bodies, they also worked in an affective register that resonates with the ways Jane Gaines talks about the phenomenology of documentary images, in which the recording and presentation of injustice prompts a shocked bodily reaction. Here "shock" designates both the physiological response to encountering an image and an alienating process that forces the body to engage other moments of historical, social, or aesthetic rupture. For more on documentary images, see Jane Gaines, "Political Mimesis," in *Collecting Visible Evidence*, ed. Jane Gaines and Michael Renov (Minneapolis: University of Minnesota Press, 1999), 84–102. For more on shock, see

Adam Lowenstein, *Shocking Representation: Historical Trauma, National Cinema, and the Modern Horror Film* (New York: Columbia University Press, 2008).

8. For more information on the spread of these images for the purposes of activist mobilization, see Yarmiar Bonilla and Jonathan Rosa, "#Ferguson: Digital Protest, Hashtag Ethnography, and the Racial Politics of Social Media in the United States," *American Ethnologist* 42, no. 1 (2015): 4–17. For more information about how information networks have facilitated the spread of politicized, traumatic imagery, see W. J. T. Mitchell, *Cloning Terror: The War of Images, 9/11 to the Present* (Chicago: University of Chicago Press, 2011).

9. Sarah Brayne, *Predict and Surveil: Data Discretion, and the Future of Policing* (New York: Oxford University Press, 2020).

10. For more on this, see Andrew Guthrie Ferguson, *The Rise of Big Data Policing: Surveillance, Race, and the Future of Law Enforcement* (New York: New York University Press, 2017).

11. Carrie Dan and Andrew Rafferty, "Obama Requests $263 Million for Body Cameras, Training," *NBC News*, December 1, 2014, www.nbcnews.com/politics/first-read/Obama-requests-263-million-police-body-cameras-training-n259161.

12. Mark Berman, "Justice Dept. Will Spend $20 Million on Police Body Cameras Nationwide," *Washington Post*, May 1, 2015, www.washingtonpost.com/news/post-nation/wp/2015/05/01/justice-dept-to-help-police-agencies-across-the-country-get-body-cameras/.

13. Dara Lind, "Obama Wants to Put Body Cameras on 50,000 More Cops," *Vox*, December 1, 2014, www.vox.com/2014/12/1/7314603/body-cameras-police.

14. Bruer, "Cameras Are No 'Panacea.'"

15. Emily Schultheis, "Hillary Clinton Calls for Body Cameras for All Police Officers Nationwide," *Atlantic*, April 29, 2015, www.theatlantic.com/politics/archive/2015/04/hillary-clinton-calls-for-body-cameras-for-all-police-officers-nationwide/457815/.

16. For one example, see Bill Nichols, *Representing Reality: Issues and Concepts in Documentary* (Bloomington: Indiana University Press, 1991).

17. Darren Palmer, "The Mythical Properties of Police Body-Worn Cameras: A Solution in the Search of a Problem," *Surveillance & Society* 14, no. 1 (2016): 138–144.

18. For example, Mary D. Fan, "Privacy, Public Disclosure, Police Body Cameras: Policy Splits," *Alabama Law Review* 68, no. 2 (2016): 337–394.

19. For example, Jack Greiner and Darren Ford, "Public Access to Police Body Camera Footage—It's Still Not Crystal Clear," *University of Cincinnati Law Review* 86, no. 1 (2018): 139–152.

20. For example, James E. Wright II and Andrea M. Headley, "Can Technology Work for Policing? Citizen Perceptions of Police-Body Worn Cameras," *American Review of Public Administration* 51, no. 1 (2021): 17–27.

21. Emmeline Taylor, "Lights, Camera, Redaction . . . Police Body-Worn Cameras: Autonomy, Discretion and Accountability," *Surveillance & Society* 14, no. 1 (2016): 128–132.

22. Stacey E. Wood, "Police Body Cameras and Professional Responsibility: Public Records and Private Evidence," *Preservation, Digital Technology, and Culture* 46, no. 1 (2017): 43.

23. Chris Pagliarella, "Police Body-Worn Camera Footage: A Question of Access," *Yale Policy & Law Review* 34 (2016): 532–543.

24. See "Police Body Cameras: Money for Nothing?" *Fox News* October 21, 2017, www.foxnews.com/politics/police-body-cameras-money-for-nothing; and German Lopez, "The Failure of Police Body Cameras," *Vox*, July 21, 2017, www.vox.com/policy-and-politics/2017/7/21/15983842/police-body-cameras-failures.

25. Radley Balko, "The Ongoing Problem of Conveniently Malfunctioning Police Cameras," *Washington Post*, June 28, 2018, www.washingtonpost.com/news/the-watch/wp/2018/06/28/the-ongoing-problem-of-conveniently-malfunctioning-police-cameras/.

26. Lily Hay Newman, "Police Bodycams Can Be Hacked to Doctor Footage," *Wired*, August 11, 2018, www.wired.com/story/police-body-camera-vulnerabilities.

27. Amanda Ripley, "A Big Test of Police Body Cameras Defies Expectations," *New York Times*, October 20, 2017, www.nytimes.com/2017/10/20/upshot/a-big-test-of-police-body-cameras-defies-expectations.html.

28. "GovDirect Rolls Out Body Worn Cameras to Greenville County Sheriff's," GovDirect, March 10, 2017, www.govdirect.com/blog/govdirect-rolls-out-body-worn-cameras-to-greenville-county-sheriffs.

29. For more on these features and to see footage of the camera in action, see Panasonic Business Solutions, "NY1 looks at the Panasonic Arbitrator BWC," posted March 8, 2017, YouTube video, www.youtube.com/watch?v=tm8Pn4c3YZ4; and Jenne, Inc., "Panasonic i-PRO Body-Worn Cameras," posted September 21, 2020, YouTube video, www.youtube.com/watch?v=PZwM_u7fiDI.

30. To see more of the camera's design and interface, see Panasonic Business Solutions, "Panasonic Arbitrator BWC," posted March 9, 2017, YouTube video, www.youtube.com/watch?v=NVa7SJaRkLI.

31. Axon, "Evolution of the Axon Body Camera," April 11, 2023, www.axon.com/blog/evolution-of-axon-body-camera.

32. Axon, "Axon Body Camera."

33. For an overview of the town's renewal from the perspective of the local government, see "Downtown Reborn," City of Greenville, accessed July 2, 2024, https://citygis.greenvillesc.gov/downtownreborn/index.html.

34. Romando Dixson, "Cost, Privacy among Concerns with Police Body Cameras," *Greenville News*, December 27, 2014, www.greenvilleonline.com/story/news/local/2014/12/27/cost-privacy-among-concerns-police-body-cameras/20938753/.

35. Meg Kinnard, "Officers Concerned about Bill Mandating SC Cops Wear Cameras," *Greenville News*, March 4, 2015, www.greenvilleonline.com/story/news/local/2015/03/04/officers-concerned-bill-mandating-sc-cops-wear-cameras/24379931/.

36. John Hopkins III, "Letter: Shooting Shows Need for Cameras," *Greenville News*, April 18, 2015, www.greenvilleonline.com/story/opinion/readers/2015/04/18/letter-police-shooting-shows-need-police-cameras/25961829/.

37. Hopkins, "Letter."

38. Michael Burns, "Greer Police Use Body Cameras but Don't Want Mandate," *Greenville News*, March 11, 2015, www.greenvilleonline.com/story/news/local/greer/2015/03/11/greer-police-use-body-cameras-want-mandate/70165008/.

39. Burns, "Police Use Body Cameras."

40. Burns, "Police Use Body Cameras."

41. Kinnard, "Officers Concerned about Bill."

42. "Editorial: Move forward on Body Cameras," *Greenville News*, April 23, 2015, www.greenvilleonline.com/story/opinion/editorials/2015/04/23/editorial-move-forward-body-cameras/26256965/.

43. Associated Press, "Haley to Sign Body Camera Legislation," *Greenville News*, June 10, 2015, www.greenvilleonline.com/story/news/politics/2015/06/10/haley-sign-body-camera-legislation-north-charleston/71009462/.

44. Mary Troyan, "Scott Wants $500 Million for Police Body Cameras," *Greenville News*, July 18, 2015, www.greenvilleonline.com/story/news/local/2015/07/28/tim-scott-police-body-cameras/30795815/.

45. Romando Dixson, "Police Discuss Body Cameras with Public," *Greenville News*, December 17, 2015, www.greenvilleonline.com/story/news/2015/12/17/police-discuss-body-cameras-public/77498962/.

46. Meg Kinnard, "SC Body Camera Law Spurred by Ferguson, State's Own Shooting," *Greenville News*, August 2, 2015, www.greenvilleonline.com/story/news/local/south-carolina/2015/08/02/sc-body-camera-law-spurred-by-ferguson-states-own-shooting/31023099/.

47. Tim Smith, "Policy Body Cam Rules Detail Who Should Wear Them and When," *Greenville New*, December 4, 2015, www.greenvilleonline.com/story/news/2015/12/04/body-camera-guidelines-approved-sc-law-enforcement/76777168/.

48. Smith, "Police Body Cam Rules."

49. Smith, "Police Body Cam Rules."

50. Dixson, "Police Discuss Body Cameras."

51. For an overview of these policies, see "Body-Worn Cameras Project," Greenville Police Department, accessed January 20, 2020, www.greenvillesc.gov/1180/Body-Worn-Cameras-Project.

52. Romando Dixson, "Greenville County Sheriff Waits on State Funds for Body Cams," *Greenville News*, June 10, 2016, www.greenvilleonline.com/story/news/crime/2016/06/10/greenville-county-sheriffs-office-sc-body-cameras/84768308/.

53. Dixson, "Sheriff Waits on State Funds."

54. Eric Connor, "Body Cameras for Greenville Police Officers Coming by End of Year," *Greenville News*, July 18, 2016, www.greenvilleonline.com/story/news/2016/07/18/body-cameras-greenville-police-officers-coming-end-year/87269504/.

55. Mike Ellis, "Only a Fraction of S.C. Police Shootings Caught on Police Video," *Greenville News*, August 19, 2016, www.greenvilleonline.com/story/news/local/south-carolina/2016/08/19/only-fraction-sc-police-shootings-caught-police-video/88990742/.

56. Romando Dixson, "Witness to Contentious Arrest: Body Cams Needed," *Greenville News,* August 3, 2016, www.greenvilleonline.com/story/news/crime/2016/08/03/greenville-county-body-cameras-arrest/87579740/.

57. In infrastructure studies, the concepts of "maintenance" and "repair" are often used to discuss how any kind of material infrastructure is not simply built, but must be actively tended to over time to avoid disruption and breakdown. Steven J. Jackson, "Rethinking Repair," in *Media Technologies: Essays on Communication, Materiality, and Society,* ed. Tarlton Gillespie, P. J. Boczkowski, and Kevin Foot (Cambridge, MA: MIT Press, 2014), 221–240.

58. Amanda Coyne, "Greenville County Gets $135k for 125 Body Cameras," *Greenville News,* October 18, 2016, www.greenvilleonline.com/story/news/2016/10/18/greenville-county-gets-135k-125-body-cameras/92362460/.

59. Tesalon Felicien, "Deputies to Get Body Cameras in Greenville This Week," *Greenville News,* March 14, 2017, www.greenvilleonline.com/story/news/crime/2017/03/14/deputies-get-body-cameras-greenville-week/99013444/.

60. Felicien, "Deputies Get Body Cameras."

61. Felicien, "Deputies Get Body Cameras."

62. These claims echo much of Steve Mann's ongoing conceptualization of veillance and the relationships between *surveillance* and *sousveillance* as they relate to wearable technologies. Mann, "New Media and Power Politics."

63. Tesalon Felicien, "Greenville Police to Wear Body Cameras by the First Week of May," *Greenville News,* April 12, 2017, www.greenvilleonline.com/story/news/crime/2017/04/12/greenville-police-wear-body-cameras-first-week-may/100365328/.

64. Felicien, "Greenville Police Wear Cameras."

65. These connections between surveillance and racism have been discussed in a good deal of surveillance studies literature. Simone Browne, *Dark Matters: On the Surveillance of Blackness* (Durham, NC: Duke University Press, 2015) is exemplary in this regard.

66. Felicien, "Greenville Police Wear Cameras."

67. Daniel J. Gross, "As Body Cameras Grow in Popularity, Greenville Police Vehicles Move Away from Dashcams," *Greenville News,* June 26, 2018, www.greenvilleonline.com/story/news/crime/2018/06/26/greenville-police-phase-out-dashcams-rely-body-cameras/714662002/.

68. Gross, "Cameras Grow in Popularity."

69. Associated Press, "Greenville County to Release Police Footage in Shootings," WLOS, November 5, 2018, https://wlos.com/news/local/greenville-county-to-release-police-footage-in-shootings.

70. These briefings are housed at "Critical Incident Videos," Los Angeles Police Department, accessed November 15, 2019, www.lapdonline.org/office-of-the-chief-of-police/professional-standards-bureau/critical-incident-videos/.

71. Daniel Grinberg, "Tracking Movements: Black Activism, Aerial Surveillance, and Transparency Optics," *Media, Culture & Society* 41, no. 3 (2019): 294–316.

72. Daniel J. Gross, "Public Won't See Bodycam video in First Greenville Deputy Shooting under New Program," *Greenville News*, March 13, 2019, www.greenvilleonline.com/story/news/local/south-carolina/2019/03/13/greenville-county-body-cameras/3137811002/.

73. Gross, "Public Won't See Bodycam."

74. Gross, "Public Won't See Bodycam."

75. Daniel J. Gross, "'We Don't Want to Kill You': Video Shows Tense Moments in Fatal Deputy-Involved Shooting," *Greenville News*, March 25, 2019, www.greenvilleonline.com/story/news/local/south-carolina/2019/03/25/sheriff-deputy-bodycam-shows-tense-moments-fatal-greenville-sc-shooting/3265979002/.

76. See James N. Gilmore, "Deathlogging: GoPros as Forensic Media in Accidental Sporting Deaths," *Convergence: The International Journal of Research into New Media Technologies* 29, no. 2 (2023): 481–495.

77. Greenville County Sheriff's Office-SC, "Greenville County Sheriff's Office CICB-2019-01," posted March 25, 2019, YouTube video, www.youtube.com/watch?v=fPMhnP83LeY.

78. Daniel J. Gross, "SC Judges Crack Down on Public Access to Police Body Camera Footage in Greenville," *Greenville News*, June 26, 2019, www.greenvilleonline.com/story/news/local/south-carolina/2019/06/26/greenville-county-sc-judges-crack-down-public-access-police-body-camera/1545756001/.

79. Gross, "SC Judges Crack Down."

80. Gross, "SC Judges Crack Down."

81. Daniel J. Gross and Haley Walters, "Greenville, Pickens Body Camera Rule Preventing Lawyers from Sharing Footage Overturned," *Greenville News*, June 27, 2019, www.greenvilleonline.com/story/news/local/south-carolina/2019/06/27/body-camera-order-greenville-pickens-vacated-sc-chief-justice/1586327001/.

82. Gross and Walters, "Greenville Body Camera Rule."

83. Gross and Walters, "Greenville Body Camera Rule."

84. Daniel J. Cross, "Greenville County Sheriff's Office Leads SC in Shootings by Law Enforcement," *Greenville News*, September 29, 2019, www.greenvilleonline.com/in-depth/news/local/south-carolina/2019/09/29/police-shootings-in-sc-greenville-county-sheriffs-office-tops-list/779095002/; Daniel J. Gross, "As Much Stress as We Put Them under, Officer Training Can't Totally Prepare for Real Crisis," *Greenville News*, September 29, 2019, www.greenvilleonline.com/in-depth/news/local/south-carolina/2019/09/29/sc-police-shootings-training-cant-totally-prepare-officers-for-crisis/3541792002/; and Daniel J. Gross, "Citizens Review for Deputy Shootings? 'It Won't Happen' in Greenville County, Sheriff Says," *Greenville News*, September 29, 2019, www.greenvilleonline.com/in-depth/news/local/south-carolina/2019/09/29/police-accountability-citizens-board-wont-happen-greenville-county/3541936002/.

85. Gross, "Citizens Review Deputy Shootings?"

86. Daniel J. Gross, "Some Police Shootings in South Carolina Aren't Captured on Body Camera: Here's Why," *Greenville News*, September 29, 2019, www

.greenvilleonline.com/in-depth/news/local/south-carolina/2019/09/29/body-camera-use-sc-police-shootings-is-inconsistent-heres-why/3210213002/.

87. Daniel J. Gross, "Secrecy of Police Body Camera Footage in SC Compounds Accountability Issue," *Greenville News*, August 3, 2020, www.greenvilleonline.com/story/news/local/south-carolina/2020/08/03/secrecy-police-body-camera-footage-compounds-accountability-issues/5553837002/.

88. Carol Motsinger and Daniel J. Gross, "SC Fails to Abide by Spirit of Its Own Law as Police Videos Spur Demands for Justice," *Greenville News*, August 3, 2020, www.greenvilleonline.com/story/news/local/south-carolina/2020/08/03/body-cameras-funding-oversight-lacking-5-years-after-new-law-sc/5341127002/.

89. Motsinger and Gross, "SC Fails."

90. Angelia L. Davis, "'We Are Tired of Marching': Greenville Activist Asks State Leaders to Enact New Laws," *Greenville News*, May 26, 2021, www.greenvilleonline.com/story/news/2021/05/26/greenville-activist-asks-sc-leaders-enact-new-laws/7440033002/.

91. Ben Brucato, "Policing Made Visible: Mobile Technologies and the Importance of Point of View," *Surveillance & Society* 13, nos. 3–4 (2015): 466.

CHAPTER 6: INFRASTRUCTURE, OR: THE DATAFICATION OF DISNEY WORLD

1. Cow Missing. "MyMagic+—Walt Disney World Resort Vacation Planning Video," posted May 20, 2017, YouTube video, www.youtube.com/watch?v=HXoTV4p4p4I.

2. Jason Farman, *The Mobile Story: Narrative Practices with Locative Technologies* (New York: Routledge, 2014). For more on information that is "given off" in public, see Erving Goffman, *The Presentation of Self in Everyday Life* (New York: Anchor, 1959).

3. There are resonances here to what Shoshana Zuboff has called "surveillance capitalism," which considers how consumers utilize "free" services (in her work, this is largely reduced to Google) in exchange for allowing their behaviors to be surveilled and sold to enrich the owners of the platform. Shoshana Zuboff, *The Age of Surveillance Capitalism: The Fight for a Human Future at the New Frontier of Power* (New York: Public Affairs, 2019).

4. Aaron Shapiro, *Design, Control Predict: Logistical Governance in the Smart City* (Minneapolis: University of Minnesota Press, 2020); and Jordan Frith, "Big Data, Technical Communication, and the Smart City," *Journal of Business and Technical Communication* 31, no. 2 (2017): 168–187. See also Anthony Townsend, *Smart Cities: Big Data, Civic Hackers, and the Quest for a New Utopia* (New York: Norton, 2013); and Rob Kitchin, "The Real-Time City? Big Data and Smart Urbanism," *GeoJournal* 79 (2014): 1–14.

5. Gilles Deleuze talks about the passcode as a central feature of control societies, in that it helps regulate and direct how things—including people and goods—travel

through space. This relates to how wearable technologies are used to manage and order elements of daily life, including, here, park operations. Gilles Deleuze, "Postscript on the Societies of Control," *October* 59 (1992): 3–7.

6. For more on this disciplinary potential, see Alice Marwick, "The Public Domain: Social Surveillance in Everyday Life," *Surveillance and Society* 9, no. 4 (2012): 378–393.

7. These transactional forms of data manufacturing throughout space can be framed as a form of infrastructural politics, as discussed in Blake Hallinan and James N. Gilmore, "Infrastructural Politics amidst the Coils of Control," *Cultural Studies* 35, nos. 4–5 (2021): 617–640. Here, we suggest that infrastructural politics operate through, among other things, the establishment and condensation of *forms* of social reality. Walt Disney World acts as a testing ground for such forms.

8. Gabriel S. Huddleston, Julie C. Garlen, and Jennifer A. Sandlin, "A New Dimension of Disney Magic: MyMagic+ and Controlled Leisure," in *Disney, Culture, and Curriculum*, ed. Jennifer A. Sandlin and Julie C. Garlen (New York: Routledge, 2016), 220–232.

9. See, for some discussions of this, Margaret J. King, "Disneyland and Walt Disney World: Traditional Values in Futuristic Form," *Journal of Popular Culture* 15, no. 1 (1981): 116–140; Christophe Bruchansky, "The Heterotopia of Disney World," *Philosophy Now* 77 (2010): 15–17; and William T. Barrie, "Disneyland and Disney World: Designing and Prescribing the Recreational Experience," *Society and Leisure* 22, no. 1 (1999): 71–82.

10. Miodrag Mitrasinovic, *Total Landscape, Theme Parks, Public Space* (New York: Routledge, 2006).

11. Bonnie J. Dow, "Vinyl Leaves: Walt Disney World and America," *Women's Studies in Communication* 19, no. 2 (1996): 251–267. See also David Allen, "Seeing Double: Disney's Wilderness Lodge," *European Journal of American Culture* 31, no. 2 (2012): 123–144; and Virginia A. Salamone and Frank A. Salamone, "Images of Main Street: Disney World and the American Adventure," *Journal of American Culture* 22, no. 1 (1999): 85–93.

12. David L. Pike, "The Walt Disney World Underground," *Space and Culture* 8, no. 1 (2005): 47–65.

13. Jean Baudrillard, *Simulacra and Simulation* (Ann Arbor: University of Michigan Press, 1994), 13.

14. There are resonances here with the notion of the "technological sublime." See Leo Marx, *The Machine in the Garden: Technology and the Pastoral Ideal in America* (New York: Oxford University Press, 1964); James W. Carey, "Historical Pragmatism and the Internet," *New Media and Society* 7, no. 4 (2005): 443–455; Joshua Reeves, "Automatic for the People: the Automation of Communicative Labor," *Communication and Critical/Cultural Studies* 13, no. 2 (2016): 150–165; and Rachel Plotnick, "Touch of a Button: Long-Distance Transmission, Communication, and Control at World's Fairs," *Critical Studies in Media Communication* 30, no. 1 (2013): 52–68.

15. Lisa Parks and Nicole Starosielski, introduction to *Signal Traffic: Critical Studies of Media Infrastructures*, ed. Lisa Parks and Nicole Starosielski (Urbana: University of Illinois Press, 2015): 4–5. Paul Dourish, building on Matthew Kirschenbaumn, calls for examining the "*relationship* between infrastructure and experience, with attention to the processes by which digital experiences are produced." Dourish locates the materiality of infrastructural elements as constituting not only what makes them "computationally feasible," but also what allows infrastructures to produce experiences. Paul Dourish, "Protocols, Packets, and Proximity," in *Signal Traffic: Critical Studies of Media Infrastructures*, ed. Lisa Parks and Nicole Staorsielski (Urbana: University of Illinois Press, 2015), 183–204.

16. Nicole Starosielski, *The Undersea Network* (Durham, NC: Duke University Press, 2015). See also Jeffrey Hecht, *City of Light: The Story of Fiber Optics* (New York: Oxford University Press, 2004).

17. For some examples of this, see Lisa Parks, *Cultures in Orbit: Satellites and the Televisual* (Durham, NC: Duke University Press, 2005); Brian Larkin, "Degraded Images, Distorted Sounds: Nigerian Video and the Infrastructure of Piracy," *Public Culture* 16, no. 2 (2004): 289–314; and Stephen Groening, *Cinema beyond Territory: Inflight Entertainment in Global Context* (London: BFI, 2014).

18. Shannon Mattern, "Deep Time of Media Infrastructures," in *Signal Traffic: Critical Studies of Media Infrastructures*, ed. Lisa Parks and Nicole Starosielski (Urbana: University of Illinois Press, 2015), 94–114.

19. Brian Larkin discusses the "unstable" yet progressive nature of infrastructure in Nigeria in *Signal and Noise: Media, Infrastructure, and Urban Culture in Nigeria* (Durham, NC: Duke University Press, 2008).

20. For some examples of this in the context of the early twentiethocentury US development of road infrastructure, see Henry Petroski, *The Road Taken: The History and Future of America's Infrastructure* (New York: Bloomsbury, 2016).

21. John Durham Peters, *The Marvelous Clouds: Towards an Elemental Theory of Media* (Chicago: University of Chicago Press, 2015): 4–5.

22. Peters, *Marvelous Clouds*, 15, 22.

23. Peters, *Marvelous Clouds*, 34.

24. Geoffrey C. Bowker and Susan Leigh Star, *Sorting Things Out: Classification and Its Consequences* (Cambridge, MA: MIT Press, 1999), 9.

25. Austin Carr, "The Messy Business of Reinventing Happiness," *Fast Company*, April 15, 2015, www.fastcompany.com/3044283/the-messy-business-of-reinventing-happiness#!.

26. Cliff Kuang, "Disney's $1 Billion Bet on a Magical Wristband," *Wired*, March 10, 2015. www.wired.com/2015/03/disney-magicband/.

27. Joel Santo Domingo, "Hands On: Disney MagicBands, MyMagic+ Web Service," *PC Magazine*, July 31, 2015, www.pcmag.com/article2/0,2817,2483861,00.asp.

28. Adrienne Shaw, "Encoding and Decoding Affordances: Stuart Hall and Interactive Media Technologies," *Media, Culture, and Society* 39, no. 4 (2017): 592–602.

29. For more on how smartphone applications help structure cultural relations, see Jeremy Wade Morris and Sarah Murray, *Appified: Culture in the Age of Apps* (Ann Arbor: University of Michigan Press, 2018).

30. For more on the technical operations of RFID, see Jordan Frith, *A Billion Little Pieces: RFID and Infrastructures of Identification* (Cambridge, MA: MIT Press, 2019).

31. Some technology writers have compared the Magic Band to the NSA, suggesting it affords Disney the opportunity to intimately "spy" on patrons through the data they provide, creating a "creepy" yet "convenient" experience. See Adam Clark Estes, "How I Let Disney Track My Every Move," Gizmodo, March 28, 2017, http://gizmodo.com/how-i-let-disney-track-my-every-move-1792875386; and Adam Weinstein, "Disney World Creepily Tracks Visitors NSA-Style with Magic Wristbands," *Gawker* (blog), January 2, 2014, http://gawker.com/disney-world-can-creepily-track-visitors-nsa-style-with-1493082046.

32. Carr, "Reinventing Happiness."

33. As quoted in Carr, "Reinventing Happiness."

34. Kuang, "Disney's $1 Billion Bet."

35. Carr, "Reinventing Happiness."

36. Frith, *Billion Little Pieces.*

37. Mark Andrejevic and Mark Burdon, "Defining the Sensor Society," *Television & New Media* 16, no. 1 (2015): 19–36, 20–21. See also Mark Andrejevic, *Automated Media* (New York: Routledge, 2019).

38. Disney Parks, "MyMagic+: You'll Want to Use It Everywhere," posted January 29, 2014, YouTube video, www.youtube.com/watch?v=2buVLVO-6F8.

39. Simon During shows how there is a longer history of such claims with respect to modern technology. Simon During, *Modern Enchantments: The Cultural Power of Secular Magic* (Cambridge, MA: Harvard University Press, 2004).

40. *Online Etymology Dictionary*, s.v. "band," accessed October 20, 2017, www.etymonline.com/index.php?term=bind&allowed_in_frame=0.

41. Pike, "Walt Disney World Underground."

42. Rob Kitchin and Martin Dodge, *Code/Space: Software and Everyday Life* (Cambridge, MA: MIT Press, 2014).

43. Pete Buczkowski and Hai Chu, "How Analytics Enhance the Guest Experience at Walt Disney World," *Analytics Magazine*, April 2012, 14–16.

44. Lisa Miller, "FastPass+: Everything You Need to Know about Walt Disney World's New System," *Huffington Post*, August 2, 2014, www.huffingtonpost.com/2014/05/29/fastpass-plus-disney-world_n_5374335.html.

45. Kuang, "Disney's $1 Billion Bet."

46. Buczkowski and Chu, "Analytics Enhance Guest Experience."

47. Ian Bogost, "Welcome to Dataland: Design Fiction at the Most Magical Place on Earth," Medium, July 29, 2014, https://medium.com/re-form/welcome-to-dataland-d8c06a5f3bc6. This implicitly resonates with work performed on the structures of "terms and uses" agreements, as well as lesser-known ways data is aggregated. Jonathan A. Obar and Anne Oeldorf-Hirsch, "The Biggest Lie on the

Internet: Ignoring the Privacy Policies and Terms of Service Policies of Social Networking Sites," *Information, Communication, & Society* 23, no. 1 (2020): 128–147.

48. Jonathan Cohn, "My TiVo Thinks I'm Gay: Algorithmic Culture and Its Discontents," *Television & New Media* 17, no. 8 (2016): 675–690.

49. James N. Gilmore, "To Affinity and Beyond: Clicking as Communicative Gesture on the Experimentation Platform," *Communication, Culture, and Critique* 13, no. 3 (2020): 333–348.

50. There are similarities here to David Lyon's concept of "social sorting," how surveillance is used to assign worth and analyze risk around human behavior. The data produced through Magic Band interactions provides, potentially, ways for Disney's engineers to understand what makes a patron more "valuable" (increased transactions, for instance). For more, see David Lyon, *Surveillance as Social Sorting: Privacy, Risk, and Automated Discrimination* (New York: Routledge, 2003).

51. Robert Payne, "Frictionless Sharing and Digital Promiscuity," *Communication and Critical/Cultural Studies* 11, no. 2 (2014): 85–102. For more about how the idea of "sharing" transforms through digital technology and file sharing, see Nicholas A. John, "File Sharing and the History of Computer: Or, Why File Sharing Is Called 'File Sharing,'" *Critical Studies in Media Communication 31*, no. 3 (2013): 198–211.

52. For a historical example of this, see Rachel Bowlby, *Carried Away: The Invention of Modern Shopping* (New York: Columbia University Press, 2000). Bowlby explores the rise of "instant" coffee and dinners in the 1950s as part of the acceleration of mass culture in the mid-twentieth century.

53. David Pogue, "Make Technology—and the World—Frictionless," *Scientific American*, April 1, 2012, www.scientificamerican.com/article/technologys-friction-problem/.

54. Whim TechNews, "Apple Pay—Official Announcement," posted September 16, 2014, YouTube video, https://youtu.be/jiqSZRKskmk.

55. For more on this, see Charles Acland, "The Crack in the Electric Window," *Cinema Journal* 51, no. 2 (2012): 167–171; and Andrejevic, *Automated Media*.

56. For more on the history of money as a form of media, see Lana Swartz, *New Money: How Payment Became Social Media* (New Haven, CT: Yale University Press, 2020).

57. David Gianatasio, "George Costanza Needs Google Wallet," *Ad Week*, September 20, 2011, www.adweek.com/creativity/george-costanza-needs-google-wallet-134976/.

58. See further Mary Douglas, *Purity and Danger: An Analysis of Concepts of Pollution and Taboo* (New York: Routledge, 1966), which explores how different forms of "contact" are understood to breed contamination and pollution—both biological and cultural.

59. Steven J. Jackson, "Rethinking Repair," in *Media Technologies: Essays on Communication, Materiality, and Society*, ed. Tarlton Gillespie, P. J. Boczkowski, and Kevin Foot (Cambridge, MA: MIT Press, 2014), 221–240.

60. Tom Staggs, "Taking the Disney Guest Experience to the Next Level," *Walt Disney World Report* (blog), January 7, 2013, https://disneyparks.disney.go.com/blog/2013/01/taking-the-disney-guest-experience-to-the-next-level/; Mark Hemmer, "How Disney Creates Magic with Technology," *OneFire*, October 29, 2015, http://blog.onefire.com/how-disney-creates-magic-with-technology/; and Chris Smith, "A Day out with Disney's MagicBand 2," *Wareable*, April 11, 2017, www.wareable.com/wearable-tech/disney-magicband-2–review.

61. *Merriam Webster's Online Dictionary*, s.v. "seam," accessed November 22, 2017, www.merriam-webster.com/dictionary/seam

62. Stephen Graham, *Disrupted Cities: When Infrastructure Fails* (New York: Routledge, 2010).

63. "Suture" was also a major theory in film studies in the 1970s, where it indicated an imagined bond that existed between "the spectator" and "the image," in which systems of continuity editing bound the film into a narrative whole while also encouraging spectators to, presumably, identify with and "enfold" themselves into moving images. For Disney World, then, the Magic Band may claim to remove "seams" of interaction, but in so doing it more intimately "sutures" bodies into an infrastructural space that is always tracking the park's population. For more, see Jean-Pierre Oudart, "Notes on Suture," *Screen* 18, no. 4 (1977–78): 35–47. For more on its critique, see William Rothman, "Against 'The System of the Suture,'" *Film Quarterly* 29, no. 1 (1975): 45–50. See also Thomas Elsaesser and Malte Hagener, *Film Theory: An Introduction through the Senses* (New York: Routledge, 2010).

64. Richard E. Foglesong, *Married to the Mouse: Walt Disney World and Orlando* (New Haven, CT: Yale University Press, 2003).

65. For more on this history, see Stephen M. Fjellman, *Vinyl Leaves: Walt Disney World and America* (New York: Routledge, 1992).

66. Gilles Deleuze, *Foucault* (New York: Bloomsbury, 2006), 27–28.

67. For more on the relationship between diagrams and materials, see Manuel De Landa, "Deleuze, Diagrams, and the Genesis of Form," *American Studies* 45, no. 1 (2000): 33–41.

68. Quoted in Foglesong, *Married to the Mouse*, 67.

69. "Community Standards," Celebration, Florida, accessed March 21, 2017, www.celebration.fl.us/town-hall/community-standards.

70. Andrew Ross, *The Celebration Chronicles: Life, Liberty, and the Pursuit of Property Value in Disney's New Town* (New York: Ballantine Books, 1999).

71. Steve Mannheim, *Walt Disney and the Quest for Community* (New York: Routledge, 2016).

72. Mannheim, *Quest for Community*, 7 (emphasis in original).

73. Rich Ling, *Taken-for-Grantedness: The Embedding of Mobile Communication in Society* (Cambridge, MA: MIT Press, 2012).

74. Mannheim, *Quest for Community*, 11.

75. As quoted in Foglesong, *Married to the Mouse*, 62.

76. Michel Foucault, *Security, Territory, Population: Lectures at the Collège de France 1977–1978*, trans. Graham Burchell (New York: Picador, 2008), 13.

77. Henri Lefebvre, *Toward an Architecture of Enjoyment* (Minneapolis: University of Minnesota Press, 2014).

78. Lefebvre, *Architecture of Enjoyment*, 11.

79. Adam Greenfield, *Everyware: The Dawning Age of Ubiquitous Computing* (Berkeley: New Riders, 2006), 64.

80. Anthony Townsend, *Smart Cities*, 15. For more on smart cities and governance, see Stephen Goldsmith and Susan Crawford, *The Responsive City: Engaging Communities Through Data-Smart Governance* (San Francisco: Jossey-Bass, 2014); and Mark Shepard, *Sentient City: Ubiquitous Computing, Architecture, and the Future of Urban Space* (Cambridge, MA: MIT Press, 2011).

81. For more on RFID tags in spatial practice, see N. Katherine Hayles, "RFID: Human Agency and Meaning in Information-Intensive Environments," *Theory, Culture & Society* 26, nos. 2–3 (2009): 47–72.

82. Townsend, *Smart Cities*, 24. See also Marcus Froth et al., *From Social Butterfly to Engaged Citizen: Urban Informatics, Social Media, Ubiquitous Computing, and Mobile Technology to Support Citizen Engagement* (Cambridge, MA: MIT Press, 2011).

83. Shapiro, *Design, Control, Predict*.

84. Germaine Haleguoua, "The Policy and Export of Ubiquitous Place," in *From Social Butterfly to Engaged Citizen: Urban Informatics, Social Media, Ubiquitous Computing, and Mobile Technology to Support Citizen Engagement*, ed. Marcus Froth et al. (Cambridge, MA: MIT Press, 2011), 322. For an example of the discriminatory function of this maintenance of efficiency in space, see Constance Gordon and Kyle Byron, "Sweeping the City: Infrastructure, Informality, and the Politics of Maintenance," *Cultural Studies* 35, nos. 4–5 (2021): 854–875.

85. Madelyn Rose Sanfilippo and Yan Shvartzshnaider, "Data and Privacy in a Quasi-Public Space: Disney World as a Smart City," in *Diversity, Divergence, Dialogue—16th International Conferencem iConference 2021, Proceedings*, ed. Katharina Toeppe, Hui Yan, and Samuel Kahi Chu (Cham, Switzerland: Springer, 2021), 235–250.

86. Kaitlyn Stone, "Enter the World of Yesterday, Tomorrow, and Fantasy: Walt Disney World's Creation and Its Implication on Privacy Rights under the MagicBand System," *Journal of High Technology Law* 18, no. 1 (2017): 230.

87. Alexander R. Galloway, *The Interface Effect* (Malden, MA: Polity, 2012), 13.

CONCLUSION: CULTURE, OR: ORDER AS A WAY OF LIFE

1. Jennifer Daryl Slack and J. Macgregor Wise, *Culture and Technology: A Primer*. 2nd ed. (New York: Peter Lang, 2015).

2. This has resonances with the concept *technoculture*, which has also been used to describe technological logics governing cultural practice, or how cultural life

becomes suffused with particular kinds of technologies focused on rationalization. See Constance Penley and Andrew Ross, *Technoculture* (Minneapolis: University of Minnesota Press, 1991).

3. Lawrence Grossberg, *We Gotta Get Out of This Place: Popular Conservatism and Postmodern Culture* (New York: Routledge, 1992), 54. See also Jennifer Daryl Slack, "The Theory and Method of Articulation," in *Stuart Hall: Critical Dialogues in Cultural Studies*, ed. David Morley and Kuan-Hsing Chen (New York: Routledge, 1996), 112–127.

4. Gilles Deleuze and Félix Guattari, *A Thousand Plateaus: Capitalism and Schizophrenia* (Minneapolis: University of Minnesota Press, 1987).

5. Slack and Wise, *Culture and Technology*, 157.

6. See also Slack and Wise, *Culture and Technology*, 158–159. For a more sustained discussion of the connections between Deleuze and Guattari in cultural studies, see Lawrence Grossberg, "Cultural Studies and Deleuze-Guattari, Part 1: A Polemic on Projects and Possibilities," *Cultural Studies* 28, no. 1 (2014): 1–28.

7. This framework is drawn from Blake Hallinan and James N. Gilmore, "Infrastructural politics amidst the coils of control," *Cultural Studies* 35, nos. 4–5 (2021): 617–640.

8. Antonio Gramsci, *The Prison Notebooks*, vols. 1–3 (New York: Columbia University Press, 2011).

9. Stuart Hall describes this at length in his lectures on the theoretical roots of cultural studies, in which Gramsci's theories of hegemony provide a pathway toward thinking about resistance, rather than overemphasizing power. Stuart Hall, *Cultural Studies 1983: A Theoretical History* (Durham, NC: Duke University Press, 2018).

10. Michel de Certeau, *The Practice of Everyday Life* (Berkeley: University of California Press, 1984).

11. Slack and Wise, *Culture and Technology*, 160–161. See also Stuart Hall, "On Postmodernism and Articulation: An Interview with Stuart Hall, Edited by Lawrence Grossberg," in *Stuart Hall: Critical Dialogues in Cultural Studies*, ed. David Morley and Kuan-Hsing Chen (New York: Routledge, 1996), 131–150.

12. Raymond Williams, "Culture Is Ordinary," in *The Everyday Life Reader*, ed. Ben Highmore (New York: Routledge, 2002), 91–100.

13. Williams, "Culture Is Ordinary," 93. Later in his career, Williams fleshed out this understanding into a model of dominant, residual, and emergent culture, which explains how forms of cultural practice come into being and become dominant at different points. Raymond Williams, *Marxism and Literature* (New York: Oxford University Press, 1977).

14. Theodor Adorno, *The Culture Industry: Selected Essays on Mass Culture* (New York: Routledge, 1991), 107.

15. Ted Striphas, *Algorithmic Culture before the Internet* (New York: Columbia University Press, 2023); Ted Striphas, "Known-Unknowns: Matthew Arnold, F. R. Leavis, and the Government of Culture," *Cultural Studies* 31, no. 1 (2017): 143–163; Norbert Elias, *The Civilizing Process: Sociogenetic and Psychogenetic Investigations*

(Malden, MA: Blackwell Publishers, 2000); and Blake Hallinan, "Civilizing Infrastructure," *Cultural Studies* 35, nos. 4–5 (2021): 707–727.

16. Striphas, *Algorithmic Culture.*

17. Lewis Mumford, *Technics and Civilization* (Chicago: University of Chicago Press, 1934).

18. Friedrich Kittler, *Discourse Networks 1800/1900* (Stanford, CA: Stanford University Press, 1990); and Harold Innis, *Empire and Communications* (Toronto: Dundurn Press, 1950).

19. Henri Lefebvre, *Everyday Life in the Modern World*, vol. 2, *Foundations for a Sociology of the Everyday* (New York: Verso, 2004), 98.

20. Lefebvre, *Everyday Life*, 21.

21. Rita Felski, *Doing Time: Feminist Theory and Postmodern Culture* (New York: New York University Press, 2000). For more on the idea of cultivation as care, see Ted Striphas, "Caring for Cultural Studies," *Cultural Studies* 33, no. 1 (2029): 1–18.

22. Deleuze and Guattari, *Thousand Plateaus*, 311. For more on the dialectical tensions between order and chaos in everyday affect using this framework, see Dan Hassoun and James N. Gilmore, "Drowsing: Toward a Concept of Sleepy Screen Engagement," *Communication and Critical/Cultural Studies* 14, no. 2 (2017): 103–119.

23. Felski, *Doing Time*; Michael Gardiner, *Critiques of Everyday life* (New York: Routledge, 2000); and Ben Highmore, *The Everyday Life Reader* (New York: Routledge, 2002).

24. For some sketches of this idea, see David Golumbia, "'Communication,' 'Critical,'" *Communication and Critical/Cultural Studies* 10, no. 3 (2013): 248–252; Victor Pickard, "Being Critical: Contesting Power within the Misinformation Society," *Communication and Critical/Cultural Studies* 10, no. 3 (2013): 306–311; Ted Striphas, "Keyword: Critical," *Communication and Critical/Cultural Studies* 10, no. 3 (2013): 324–328; and Sarah Banet-Weiser, "Locating Critique," *Communication and Critical/Cultural Studies* 10, no. 3 (2013): 229–232.

25. For some conceptualization of *ambivalence*, see Taina Bucher, "Bad Guys and Bag Ladies: On the Politics of Polemics and the Promise of Ambivalence," *Social Media + Society* 5, no. 3 (2019), https://doi.org/10.1177/2056305119856705.

26. Algorithmic Justice League, home page, accessed February 20, 2024, www.ajl.org; Global Indigenous Data Alliance, home page, accessed February 20, 2024, www.gida-global.org; *and* Distributed AI Research Institute, home page, accessed February 20, 2024, www.dair-institute.org.

27. Michel Foucault, *Language, Counter-Memory, Practice: Selected Essays and Interviews*, ed. Donald F. Bouchard (New York: Cornell University Press, 1977).

28. Michel Foucault, *Power/Knowledge: Selected Interviews and Other Writings, 1972–1977*, ed. Colin Gordon (New York: Pantheon, 1980), 131–132.

29. Foucault, *Power/Knowledge.*

30. Jacinthe Flore, "Ingestible Sensors, Data, and Pharmaceuticals: Subjectivity in the Era of Digital Mental Health," *New Media & Society* 23, no. 7 (2021): 2034–2051.

31. James N. Gilmore, "Deathlogging: GoPros as Forensic Media in Accidental Sporting Deaths," *Convergence* 29, no. 2 (2023): 481–495.

32. For just one example, see Olya Kudina and Peter-Paul Verbeek, "Ethics from Within: Google Glass, the Collingridge Dilemma, and the Mediated Value of Privacy," *Science, Technology, and Human Values* 44, no. 2 (2019): 291–314.

33. Raymond Williams, *The Politics of Modernism: Against the New Conformists* (New York: Verso, 1991), 134.

BIBLIOGRAPHY

Abbate, Emily. "Here's How the NBA's Coronavirus-Fighting Ring Might Help." *GQ*, June 17, 2020. www.gq.com/story/oura-ring-nba.

Acland, Charles. "The Crack in the Electric Window." *Cinema Journal* 51, no. 2 (2012): 167–171.

Adorno, Theodor. *The Culture Industry: Selected Essays on Mass Culture*. New York: Routledge, 1991.

Ahmed, Sara. *Queer Phenomenology: Orientations, Objects, Others*. Durham, NC: Duke University Press, 2006.

Ahn, Sun Joo (Grace), Kyle Johnsen, and Catherine Ball. "Points-Based Reward Systems in Gamification Impact Children's Physical Activity Strategies and Psychological Needs." *Health Education & Behavior* 46, no. 3 (2019): 417–425.

Agamben, Giorgio. *What Is an Apparatus?* Stanford, CA: Stanford University Press, 2009.

Ajana, Btihaj. *Governing through Biometrics: The Politics of Identity*. New York: Palgrave Macmillan, 2013.

Ajunwa, Ijfeoma, Kate Crawford, and Jason Schultz. "Limitless Worker Surveillance." *California Law Review* 105 (2017): 735–776.

Alabama Crimson Tide on AL.com. "Alabama Football Uses Top of the Line GPS Technology to Track Player Performance." Posted on July 2, 2017. YouTube video. www.youtube.com/watch?v=DT7Yk7ED1QM&t=29s.

Algorithmic Justice League. Home page. Accessed February 15, 2024. www.ajl.org.

Allen, David. "Seeing Double: Disney's Wilderness Lodge." *European Journal of American Culture* 31, no. 2 (2012): 123–144.

Alper, Meryl. *Giving Voice: Mobile Communication, Disability, and Inequality*. Cambridge, MA: MIT Press, 2016.

Althusser, Louis. *On the Reproduction of Capitalism*. New York: Verso, 2016.

Ames, Morgan G. *The Charisma Machine: The Life, Death, and Legacy of One Laptop per Child*. Cambridge, MA: MIT Press, 2019.

Anderson, Chris. "The End of Theory: The Data Deluge Makes the Scientific Method Obsolete." *Wired*, June 23, 2008. www.wired.com/2008/06/pb-theory/.

Andre, Rahina. "Humanyze Announces New Workplace Strategy Solution to Help Companies Inform and Improve Workplace Decisions with Science-Backed Insights." Humanyze. August 17, 2021. https://humanyze.com/humanyze-announces-new-workplace-strategy-solution-to-help-companies-inform-and-improve-workplace-decisions-with-science-backed-insights/.

Andrejevic, Mark. *Automated Media*. New York: Routledge, 2019.

———. "Surveillance in the Digital Enclosure." *Communication Review* 10, no. 4 (2007): 295–317.

Andrejevic, Mark, and Mark Burdon. "Defining the Sensor Society." *Television and New Media* 16, no. 1 (2015): 19–36.

Apple. "Apple Watch—Dear Apple—Apple." Posted on September 12, 2017. YouTube video. www.youtube.com/watch?v=N-x8Ik9G5Dg.

———. "Apple Watch/Dear Apple/Apple." Posted on September 7, 2022. YouTube video. www.youtube.com/watch?v=fOHj5kGU4fY.

———. "Nike and Apple Team up to Launch Nike+iPod." Newsroom. May 23, 2006. www.apple.com/newsroom/2006/05/23Nike-and-Apple-Team-Up-to-Launch-Nike-iPod/.

———. "Rise." Posted April 24, 2014. YouTube video. www.youtube.com/watch?v=_APfS8aoayQ.

Aquino, Steven. "AirPods to Get Live Listen Feature in iOS 12." *Tech Crunch*, June 5, 2018. https://techcrunch.com/2018/06/05/airpods-to-get-live-listen-feature-in-ios-12/.

———. "At Apple's WWDC 2018, Accessibility Pervades All." *Tech Crunch*, June 7, 2018. https://techcrunch.com/2018/06/07/at-apples-wwdc-2018-accessibility-pervades-all/.

Armstrong, David. "The Rise of Surveillance Medicine." *Sociology of Health & Illness* 17, no. 3 (1995): 393–404.

Associated Press. "Connecticut Man Sentenced to 65 Years for Wife's Killing in 'Fitbit Murder' Case." *NBC News*, August 18, 2022. www.nbcnews.com/news/us-news/connecticut-man-sentenced-65-years-wifes-killing-fitbit-murder-case-rcna43859.

———. "Greenville County to Release Police Footage in Shootings." WLOS, November 5, 2018. https://wlos.com/news/local/greenville-county-to-release-police-footage-in-shootings.

———. "Haley to Sign Body Camera Legislation." *Greenville News*, June 10, 2015. www.greenvilleonline.com/story/news/politics/2015/06/10/haley-sign-body-camera-legislation-north-charleston/71009462/.

———. "ORU Students Log 3 Billion Steps on Fitbit Devices." March 20, 2017. www.apnews.com/4574909b130d475089f6970f816d9f65.

"Athletics Partners with WHOOP to Support Student-Athlete Holistic Wellness." Penn State Athletics. August 16, 2022. https://gopsusports.com/news/2022/8/16/general-athletics-partners-with-whoop.aspx.

Axon. "Evolution of the Axon Body Camera." April 11, 2023. www.axon.com/blog/evolution-of-axon-body-camera.

Baja, Samar. "Racial Bias Is Built into the Design of Pulse Oximeters." *Washington Post*, July 27, 2022. www.washingtonpost.com/made-by-history/2022/07/27/racial-bias-is-built-into-design-pulse-oximeters/.

Balko, Radley. "The Ongoing Problem of Conveniently Malfunctioning Police Cameras." *Washington Post*, June 28, 2018. www.washingtonpost.com/news/the-watch/wp/2018/06/28/the-ongoing-problem-of-conveniently-malfunctioning-police-cameras/.

Ball, Kirstie. "Workplace Surveillance: An Overview." *Labor History* 51, no. 1 (2010): 87–106.

Balletta, John A. "Measuring Baseball's Heartbeat: The Hidden Harms of Wearable Technology to Professional Ballplayers." *Duke Law and Technology Review* 18, no. 1 (2019): 268–292.

Banet-Weiser, Sarah. "Locating Critique." *Communication and Critical/Cultural Studies* 10, no. 3 (2013): 229–232.

Barlyn, Suzanne. "Strap on the Fitbit: John Hancock to Sell Only Interactive Life Insurance." *Reuters*, September 19, 2018. www.reuters.com/article/us-manulife-financi-john-hancock-lifeins/strap-on-the-fitbit-john-hancock-to-sell-only-interactive-life-insurance-idUSKCN1LZ1WL.

Barrie, William T. "Disneyland and Disney World: Designing and Prescribing the Recreational Experience." *Society and Leisure* 22, no. 1 (1999): 71–82.

Barthes, Roland. *Camera Lucida: Reflections on Photography*. New York: Hill and Wang, 2010.

Baudrillard, Jean. *Simulacra and Simulation*. Translated by Sheila Farid Glaser. Ann Arbor: University of Michigan Press, 1994.

Baudry, Jean-Louis. "Ideological Effects of the Basic Cinematographic Apparatus." In *Film Theory & Criticism*, 6th ed., edited by Leo Braudy and Marshall Cohen, 355–365. New York: Oxford University Press, 2004.

Beamon, Krystal K. "'Used Goods': Former African American College Student-Athletes' Perception of Exploitation by Division I Universities." *Journal of Negro Education* 77, no. 4 (2008): 352–364.

Beck, Ulrich. *Risk Society: Towards a New Modernity*. London: Sage Publications, 1992.

Beer, David. *The Data Gaze: Capitalism, Power, and Perception*. Thousand Oaks, CA: Sage Publications, 2018.

BeetTV. "TechCrunch 50 Runner-Up Fitbit Tracks Activities." Posted September 12, 2008. YouTube video. www.youtube.com/watch?v=oW3JRr9PIHg.

Bellin, Jeffrey, and Shevarma Pemberton. "Policing the Admissibility of Body Camera Evidence." *Fordham Law Review* 87, no. 4 (2019): 1425–1457.

Beniger, James R. *The Control Revolution: Technological and Economic Origins of the Information Society*. Cambridge, MA: Harvard University Press, 1986.

Benjamin, Ruha. *Race after Technology: Abolitionist tools for the New Jim Code*. Malden, MA: Polity, 2019.

Berger, Peter, and Thomas Luckmann. *The Social Construction of Reality: A Treatise on the Sociology of Knowledge*. New York: Anchor Books, 1967.

Berman, Mark. "Justice Dept. Will Spend $20 Million on Police Body Cameras Nationwide." *Washington Post*, May 1, 2015. www.washingtonpost.com/news/post-nation/wp/2015/05/01/justice-dept-to-help-police-agencies-across-the-country-get-body-cameras/.

Biggs, John. "TC50: FitBit, a Fitness Gadget That Makes Us Want to Exercise." *TechCrunch*, September 9, 2008. https://techcrunch.com/2008/09/09/tc50-fitbit-fitness-gadget-the-makes-us-want-to-exercise.

Blanchot, Maurice. "Everyday Speech." *Yale French Studies* 73 (1987): 12–20.

Blinder, Alan. "Michael Slager, Officer in Walter Scott Shooting, Gets 20-Year Sentence." *New York Times*, December 7, 2017. www.nytimes.com/2017/12/07/us/michael-slager-sentence-walter-scott.html.

"Body-Worn Cameras Project." Greenville Police Department. Accessed January 20, 2020. www.greenvillesc.gov/1180/Body-Worn-Cameras-Project.

Bogost, Ian. "Welcome to Dataland: Design Fiction at the Most Magical Place on Earth." Medium. July 29, 2014. https://medium.com/re-form/welcome-to-dataland-d8c06a5f3bc6.

———. "Why Gamification Is Bullshit." In *The Gamified World: Approaches, Issues, Applications*, edited by Steffen P. Walz and Sebastian Deterding, 65–81. Cambridge, MA: MIT Press, 2014.

Bolter, Jay David, and Richard Grusin. *Remediation: Understanding New Media*. Cambridge, MA: MIT Press, 1999.

Bonilla, Yarmiar, and Jonathan Rosa, "#Ferguson: Digital Protest, Hashtag Ethnography, and the Racial Politics of Social Media in the United States." *American Ethnologist* 42, no. 1 (2015): 4–17.

Bouk, Dan. *How Our Days Became Numbered: Risk and the Rise of the Statistical Individual*. Chicago: University of Chicago Press, 2015.

Bourdieu, Pierre. *Distinction: A Social Critique of the Judgment of Taste*. Cambridge, MA: Harvard University Press, 1984.

Bowker, Geoffrey, and Susan Leigh Star. *Sorting Things Out: Classification and Its Consequences*. Cambridge, MA: MIT Press, 1999.

Bowlby, Rachel. *Carried Away: The Invention of Modern Shopping*. New York: Columbia University Press, 2000.

Brand, Stewart. *The Media Lab: Inventing the Future at MIT*. New York: Viking Books, 1987.

Brayne, Sarah. *Predict and Surveil: Data, Discretion, and the Future of Policing*. New York: Oxford University Press, 2020.

Bregman, Albert. *Auditory Science Analysis: The Perceptual Organization of Sound*. Cambridge, MA: MIT Press, 1990.

Brown, Lauren Alix. "AirPods Could Revolutionize What It Means to Be Hard of Hearing." *Quartz*, July 7, 2018. https://qz.com/1323215/apples-airpods-and-live-listen-are-a-revolution-for-the-hearing-impaired/.

Brown, Sarah M., and Katie M. Brown. "Should Your Wearables Be Shareable? The Ethics of Wearable Technology in Collegiate Athletics." *Marquette Sports Law Review* 32, no. 1 (2021): 97–116.

Browne, Simone. *Dark Matters: On the Surveillance of Blackness*. Durham, NC: Duke University Press, 2015.
Brucato, Ben. "Policing Made Visible: Mobile Technologies and the Importance of Point of View." *Surveillance & Society* 13, nos. 3–4 (2015): 455–473.
Bruchansky, Christophe. "The Heterotopia of Disney World." *Philosophy Now* 77 (2010): 15–17.
Bruer, Wesley. "Obama Warns Cop Body Cameras Are No 'Panacea.'" *CNN*, March 3, 2015. www.cnn.com/2015/03/02/politics/obama-police-body-camera-report/index.html.
Bucher, Taina. "Bad Guys and Bag Ladies: On the Politics of Polemics and the Promise of Ambivalence." *Social Media + Society* 5, no. 3 (2019). https://doi.org/10.1177/2056305119856705.
Buczkowski, Pete, and Hai Chu. "How Analytics Enhance the Guest Experience at Walt Disney World." *Analytics Magazine*, April 2012, 14–16.
Bull, Michael. *Sounding Out the City: Personal Stereos and the Management of Everyday Life*. New York: Berg, 2000.
Bunn, Geoffrey C. *The Truth Machine: A Social History of the Lie Detector*. Baltimore, MD: Johns Hopkins University Press, 2012.
Burns, Joseph E., and Paul R. Sawyer. "The Portable Music Player as a Defense Mechanism." *Journal of Radio & Audio Media* 17, no. 1 (2010): 96–108.
Burns, Michael. "Greer Police Use Body Cameras but Don't Want Mandate." *Greenville News*, March 11, 2015. www.greenvilleonline.com/story/news/local/greer/2015/03/11/greer-police-use-body-cameras-want-mandate/70165008/.
Butler, Judith. *Gender Trouble: Feminism and the Subversion of Identity*. New York: Routledge, 1992.
Butler, Sarah. "Amazon Accused of Treating UK Warehouse Staff Like Robots." *Guardian*, May 31, 2018. www.theguardian.com/business/2018/may/31/amazon-accused-of-treating-uk-warehouse-staff-like-robots.
Caldwell, John Thornton. *Production Culture: Industrial Reflexivity and Critical Practice in Film and Television*. Durham, NC: Duke University Press, 2006.
Carey, James W. "Historical Pragmatism and the Internet." *New Media and Society* 7, no. 4 (2005): 443–455.
Carman, Ashley, Sean O'Kane, Rebecca Jennings, and Cameron Wolf. "Are AirPods Fashionable?" *Verge*, January 25, 2017. www.theverge.com/circuitbreaker/2017/1/25/14384112/apple-airpods-fashion-style-wireless-earbuds.
Carr, Austin. "The Messy Business of Reinventing Happiness." *Fast Company*, April 15, 2015. www.fastcompany.com/3044283/the-messy-business-of-reinventing-happiness#!.
Catapult, "Catapult Teams Dominate College Football Standings." *PR Newswire*, December 21, 2021. www.prnewswire.com/news-releases/catapult-teams-dominate-college-football-standings-301449189.html.
Chan, Edward D., Michael M. Chan, and Mallory M. Chan. "Pulse Oximetry: Understanding Its Basic Principles Facilitates Appreciation of Its Limitations." *Respiratory Medicine* 107, no. 6 (2013): 789–799.

Chan, Rosalie. "The High-Tech Office of the Future Will Spy on You." *Week*, June 6, 2018. https://theweek.com/articles/760582/hightech-office-future-spy.

Chandler, Alfred D., Jr. *The Visible Hand: The Managerial Revolution in American Business*. Cambridge, MA: Harvard University Press, 1993.

Chariana, Shams. "Inside the NBA Bubble: Details from the NBPA Memo Obtained by The Athletic." *Athletic*, June 16, 2020. https://theathletic.com/1876737/2020/06/16/inside-the-nba-bubble-details-from-nbpa-memo-obtained-by-the-athletic.

Chen, Angela. "What the Apple Watch's FDA Clearance Actually Means." *Verge*, September 13, 2018. www.theverge.com/2018/9/13/17855006/apple-watch-series-4-ekg-fda-approved-vs-cleared-meaning-safe.

Cheney-Lippold, John. *We Are Data: Algorithms and the Making of Our Digital Selves*. New York: New York University Press, 2017.

Chernega, Jennifer. "Black Lives Matter: Racialized Policing in the United States." *Comparative American Studies* 14, nos. 3–4 (2016): 234–245.

Chuck, Elizabeth. "Oral Roberts University to Track Students' Fitness through Fitbits." *NBC News*, February 3, 2016. www.nbcnews.com/feature/college-game-plan/oral-roberts-university-track-students-fitness-through-fitbits-n507661.

Chun, Wendy Hui Kyong. *Updating to Remain the Same: Habitual New Media*. Cambridge, MA: MIT Press, 2016.

Clarke, Roger. "The Digital Persona and Its Application to Data Surveillance." *Information Society* 10, no. 2 (1994): 77–92.

CNET. "Here One 'Smart' Wireless Earphones Aren't AirPod Killers, but They're Better in Some Ways." Posted February 21, 2017. YouTube video. www.youtube.com/watch?v=XoJWDcSzwsI.

Cohen, B. R. "Public Thinker: Jill Lepore on the Challenge of Explaining Things." *Public Books*, April 24, 2017. www.publicbooks.org/public-thinker-jill-lepore-on-the-challenge-of-explaining-things.

Cohn, Jonathan. *The Burden of Choice: Recommendations, Subversion, and Algorithmic Culture*. New Brunswick, NJ: Rutgers University Press, 2019.

———. "My TiVo Thinks I'm Gay: Algorithmic Culture and Its Discontents." *Television & New Media* 17, no. 8 (2016): 675–690.

Cohn, Jonathan Evan. "Ultrasonic Bracelet and Receiver for Detecting Position in 2D Plane." Google Patents. Accessed May 20, 2018. https://patents.google.com/patent/US20170278051A1/en.

"Community Standards." Celebration, Florida. Accessed March 21, 2017. www.celebration.fl.us/town-hall/community-standards.

Comstock, Jonah. "Fitbit CEO Hints at Expanding Healthcare Strategy, FDA-Cleared Devices." *Mobi Health News*, May 5, 2016. www.mobihealthnews.com/content/fitbit-ceo-hints-expanding-healthcare-strategy-fda-cleared-devices.

———. "VA to Reimburse for Certain Clinical Activity Tracks." *Mobi Health News*, August 28, 2014. www.mobihealthnews.com/36158/va-to-reimburse-for-certain-clinical-activity-trackers.

Connor, Eric. "Body Cameras for Greenville Police Officers Coming by End of Year." *Greenville News*, July 18, 2016. www.greenvilleonline.com/story/news/2016/07/18/body-cameras-greenville-police-officers-coming-end-year/87269504/.

"Consumers File Class Action against Fitbit, Inc. Alleging That Fitbit Charge HR and Surge Heart Rate Monitors Do Not Accurately Track Users' Heart Rate." *Business Wire*, January 5, 2016. www.businesswire.com/news/home/20160105006634/en/Consumers-File-Class-Action-Against-Fitbit-Inc.-Alleging-That-Fitbit-Charge-HR-and-Surge-Heart-Rate-Monitors-Do-Not-Accurately-Track-Users'-Heart-Rates.

Cooley, Heidi Rae. *Finding Augusta: Habits of Mobility and Governance in the Digital Era*. Hanover, NH: Dartmouth College Press, 2014.

Cow Missing. "MyMagic+—Walt Disney World Resort Vacation Planning Video." Posted May 20, 2017. YouTube video. www.youtube.com/watch?v=HXoTV4p4p4I.

Cowan, Ruth Schwartz. *More Work for Mother: The Ironies of Household Technology from the Open Hearth to the Microwave*. New York: Basic Books, 1985.

Coyne, Amanda. "Greenville County Gets $135k for 125 Body Cameras." *Greenville News*, October 18, 2016. www.greenvilleonline.com/story/news/2016/10/18/greenville-county-gets-135k-125-body-cameras/92362460/.

Crawford, Kate. *Atlas of AI: Power, Politics, and the Planetary Costs of Artificial Intelligence*. New Haven, CT: Yale University Press, 2021.

"Critical Incident Videos." Los Angeles Police Department. Accessed November 15, 2019. www.lapdonline.org/office-of-the-chief-of-police/professional-standards-bureau/critical-incident-videos/.

Daily Collegian, The. "WHOOP There It Is: James Franklin Explains Penn State Football's Wearable Sleep Monitor." Posted September 13, 2017. YouTube video. www.youtube.com/watch?v=FmcScmytHg8.

Dan, Carrie, and Andrew Rafferty. "Obama Requests $263 Million for Body Cameras, Training." *NBC News*, December 1, 2014. www.nbcnews.com/politics/first-read/Obama-requests-263-million-police-body-cameras-training-n259161.

Davenport, Thomas, and John Beck. *The Attention Economy: Understanding the New Currency: of Business*. Cambridge, MA: Harvard Business School Press, 2001.

Davis, Angela L. "'We Are Tired of Marching': Greenville Activist Asks State Leaders to Enact New Laws." *Greenville News*, May 26, 2021. www.greenvilleonline.com/story/news/2021/05/26/greenville-activist-asks-sc-leaders-enact-new-laws/7440033002/.

Davis, Lennard. *Enforcing Normalcy: Disability, Deafness, and the Body*. New York: Verso, 1995.

De Certeau, Michel. *The Practice of Everyday Life*. Berkeley: University of California Press, 1984.

De Landa, Manuel. "Deleuze, Diagrams, and the Genesis of Form," *American Studies* 45, no. 1 (2000): 33–41.

De Montjoyve, Yves-Alexandre, Laura Radaeleli, Vivek Kumar Singh, and Alex 'Sandy' Pentland. "Unique in the Shopping Mall: On the Reidentifiability of Credit Card Metadata," *Science* 347, no. 6221 (2015): 536–539.

Deleuze, Gilles. *Difference and Repetition*. New York: Columbia University Press, 1995.

———. *Foucault*. New York: Bloomsbury, 2006.

———. "Postscript on the Societies of Control." *October* 59 (1992): 3–7.

Deleuze, Gilles, and Félix Guattari. *A Thousand Plateaus: Capitalism and Schizophrenia*. Minneapolis: University of Minnesota Press, 1987.

Delfanti, Alessandro. "Machinic Dispossession and Augmented Despotism: Digital Work in an Amazon Warehouse." *New Media & Society* 23, no. 1(2019): 39–55.

DeNora, Tia. "Music as a Technology of the Self." *Poetics* 27, no. 1 (1999): 31–56.

"Design for Everyone." Apple Developer. 2017. https://developer.apple.com/videos/play/wwdc2017/806/. Accessed at https://web.archive.org/web/20181010035537/https:/developer.apple.com/videos/play/wwdc2017/806/.

DiBenedetto, Chase. "The Apple Watch's Blood Oxygen Measurement Might Be Guilty of Racial Bias." *Mashable*, December 28, 2022. https://mashable.com/article/apple-watch-oximeter-racial-bias-lawsuit.

Disney Parks. "MyMagic+: You'll Want to Use it Everywhere." Posted January 29, 2014. YouTube video. www.youtube.com/watch?v=2buVLVO-6F8.

Distributed AI Research Institute. Home page. Accessed February 20, 2024. www.dair-institute.org.

Dixson, Romando. "Cost, Privacy among Concerns with Police Body Cameras." *Greenville News*, December 27, 2014. www.greenvilleonline.com/story/news/local/2014/12/27/cost-privacy-among-concerns-police-body-cameras/20938753/.

———. "Greenville County Sheriff Waits on State Funds for Body Cams." *Greenville News*, June 10, 2016. www.greenvilleonline.com/story/news/crime/2016/06/10/greenville-county-sheriffs-office-sc-body-cameras/84768308/.

———. "Police Discuss Body Cameras with Public." *Greenville News*, December 17, 2015. www.greenvilleonline.com/story/news/2015/12/17/police-discuss-body-cameras-public/77498962/.

———. "Witness to Contentious Arrest: Body Cams Needed." *Greenville News*, August 3, 2016. www.greenvilleonline.com/story/news/crime/2016/08/03/greenville-county-body-cameras-arrest/87579740/.

Domingo, Joel Santo. "Hands On: Disney MagicBands, MyMagic+ Web Service." *PC Magazine*, July 31, 2015. www.pcmag.com/article2/0,2817,2483861,00.asp.

Douglas, Mary. *Purity and Danger: An Analysis of Concepts of Pollution and Taboo*. New York: Routledge, 1966.

Dourish, Paul. "Protocols, Packets, and Proximity." In *Signal Traffic: Critical Studies of Media Infrastructures*, edited by Lisa Parks and Nicole Staorsielski, 183–204. Urbana: University of Illinois Press, 2015.

Dow, Bonnie J. "Vinyl Leaves: Walt Disney World and America." *Women's Studies in Communication* 19, no. 2 (1996): 251–267.

"Downtown Reborn." City of Greenville. Accessed July 2, 2024. https://citygis.greenvillesc.gov/downtownreborn/index.html.

During, Simon. *Modern Enchantments: The Cultural Power of Secular Magic.* Cambridge, MA: Harvard University Press, 2004.

Dzieza, Josh. "'Beat the Machine': Amazon Warehouse Workers Strike to Protest Inhumane Conditions." *Verge*, July 16, 2019. www.theverge.com/2019/7/16/20696154/amazon-prime-day-2019-strike-warehouse-workers-inhumane-conditions-the-rate-productivity.

Eadicicco, Lisa. "Smartwatches Have Measured Blood Oxygen for Years: But Is This Useful?" *CNet*, June 16, 2022. www.cnet.com/tech/mobile/smartwatches-have-measured-blood-oxygen-for-years-but-is-it-useful/.

"Earn Vitality Points for Healthy Living." John Hancock/Vitality. Accessed September 15, 2019. https://jh1.jhlifeinsurance.com/jhl-ext-templating/filedetail?vgnextoid=83b97a0fbe21c410VgnVCM1000003e86fa0aRCRD&siteName=JHSalesNet.

"Editorial: Move Forward on Body Cameras." *Greenville News*, April 23, 2015. www.greenvilleonline.com/story/opinion/editorials/2015/04/23/editorial-move-forward-body-cameras/26256965/.

Eichner, Randy E. "Football Team Rhabdomyolysis: The Pain Beats the Gain and the Coach Is to Blame." *Current Sports Medicine Reports* 17, no. 5 (2018): 142–143.

Eisenberg, Anne. "The Hearing Aid as Fashion Statement." *New York Times,* September 24, 2006. www.nytimes.com/2006/09/24/business/yourmoney/24novel.html.

Eley, Louise, and Ben Rampton. "Everyday Surveillance, Goffman, and Unfocused Interaction." *Surveillance & Society* 18, no. 2 (2020): 199–215.

Elias, Norbert. *The Civilizing Process: Sociogenetic and Psychogenetic Investigations.* Malden, MA: Blackwell Publishers, 2000.

Ellcessor, Elizabeth. *Restricted Access: Media, Disability, and the Politics of Participation.* New York: New York University Press, 2016.

Ellis, Katie, and Mike Kent. *Disability and New Media.* New York: Routledge, 2011.

Ellis, Mike. "Only a Fraction of S.C. Police Shootings Caught on Police video." *Greenville News*, August 19, 2016. www.greenvilleonline.com/story/news/local/south-carolina/2016/08/19/only-fraction-sc-police-shootings-caught-police-video/88990742/.

Ellul, Jacques. *The Technological Society.* New York: Vintage Books, 1964.

Elman, Julie Passanante. "'Find Your Fit': Wearable Technology and the Cultural Politics of Disability." *New Media & Society* 20, no. 10 (2018): 3760–3777.

Elsaesser, Thomas, and Malte Hagener. *Film Theory: An Introduction through the Senses.* New York: Routledge, 2010.

Emanuel, Ezekiel, Aaron Glickman, and David Johnson. "Measuring the Burden of Health Care Costs on US Families: The Affordability Index." *JAMA* 318, no. 19 (2017): 1863–1864.

Erlmann, Veit. "But What of the Ethnographic Ear? Anthropology, Sound, and the Senses." In *Hearing Cultures: Essays on Sound, Listening, and Modernity*, edited by Veit Erlmann, 1–20. New York: Bloomsbury, 2004.

Estes, Adam Clark. "How I Let Disney Track My Every Move." Gizmodo. March 28, 2017. http://gizmodo.com/how-i-let-disney-track-my-every-move-1792875386.

Etherington, Darrel. "Researchers Use Biometrics, Including Data from the Oura Ring, to Predict COVID-19 Symptoms in Advance." *Tech Crunch*, May 28, 2020. https://techcrunch.com/2020/05/28/researchers-use-biometrics-including-data-from-the-oura-ring-to-predict-covid-19-symptoms-in-advance/.

Eubanks, Virginia. *Automating Inequality: How High-Tech Tools Profile, Police, and Punish the Poor*. New York: St. Martin's Press, 2018.

Ewen, Stuart. *Captains of Consciousness: Advertising and the Social Roots of Consumer Culture*. New York: Basic Books, 1976.

Fairclough, Norman. *Critical Discourse Analysis: The Critical Study of Language*. New York: Routledge, 1995.

Falter, Maarten, Werner Budts, Katie Goetschalckx, Veronique Cornelissen, and Roselien Buys. "Accuracy of Apple Watch Measurements for Heart Rate and Energy Expenditure in Patients with Cardiovascular Disease: Cross-Sectional Study." *JMIR Mhealth Uhealth* 7, no. 3 (2019): e11889.

Fan, Mary D. "Privacy, Public Disclosure, Police Body Cameras: Policy Splits." *Alabama Law Review* 68, no. 2 (2016): 337–394.

Farman, Jason. *The Mobile Story: Narrative Practices with Locative Technologies*. New York: Routledge, 2014.

Farr, Christina. "Apple Heart Study Shows a Lot of Promise for Digital Health, but Cardiologists Still Have Questions." *CNBC*, November 13, 2019. www.cnbc.com/2019/11/13/apple-releases-heart-study-results.html.

Felicien, Tesalon. "Deputies to Get Body Cameras in Greenville This Week." *Greenville News*, March 14, 2017. www.greenvilleonline.com/story/news/crime/2017/03/14/deputies-get-body-cameras-greenville-week/99013444/.

———. "Greenville Police to Wear Body Cameras by the First Week of May." *Greenville News*, April 12, 2017. www.greenvilleonline.com/story/news/crime/2017/04/12/greenville-police-wear-body-cameras-first-week-may/100365328/.

Felski, Rita. *Doing Time: Feminist Theory and Postmodern Culture*. New York: New York University Press, 2000.

Ferguson, Andrew Guthrie. *The Rise of Big Data Policing: Surveillance, Race, and the Future of Law Enforcement*. New York: New York University Press, 2017.

Ferreira, João J., Cristina I. Fernandes, Hussain G. Rammal, and Pedo M. Veiga. "Wearable Technology and Consumer Interaction: A Systematic Review and Research Agenda." *Computers in Human Behavior* 118 (2021): 1–10.

Finn, Ed. *What Algorithms Want: Imagination in the Age of Computing*. Cambridge, MA: MIT Press, 2018.

Fisher, Christine. "Researchers Say Oura Rings Can Predict COVID-19 Symptoms Three Days Early." *Engadget*, June 1, 2020. www.engadget.com/west-virginia-university-oura-ring-covid-19-symptoms-003239603.html.

"Fitbit Charge HR TV Commercial, 'Know Your Heart.'" iSpot. Accessed May 30, 2022. www.ispot.tv/ad/7QpS/fitbit-charge-hr-know-your-heart.

"Fitbit Charge 2—Big Day." Posted April 5, 2017. YouTube video. www.youtube.com/watch?v=bW5ihFg-5iQ.
"Fitbit Corporate Wellness Solutions." Fitbit. Accessed September 8, 2019. www.fitbit.com/en-ca/product/corporate-solutions
"Fitbit Launches Fitbit Care, a Powerful New Enterprise Health Platform for Wellness and Prevention and Disease Management." Fitbit. September 18, 2018. https://investor.fitbit.com/press/press-releases/press-release-details/2018/Fitbit-Launches-Fitbit-Care-A-Powerful-New-Enterprise-Health-Platform-for-Wellness-and-Prevention-and-Disease-Management/default.aspx.
Fjellman, Stephen M. *Vinyl Leaves: Walt Disney World and America*. New York: Routledge, 1992.
Flore, Jacinthe. "Ingestible Sensors, Data, and Pharmaceuticals: Subjectivity in the Era of Digital Mental Health." *New Media & Society* 23, no. 7 (2021): 2034–2051.
Foglesong, Richard E. *Married to the Mouse: Walt Disney World and Orlando*. New Haven, CT: Yale University Press, 2003,
Food and Drug Administration. "De Novo Classification Request for ECG App." August 14, 2018. www.accessdata.fda.gov/cdrh_docs/reviews/DEN180044.pdf.
Fotopoulou, Artistes, and Kate O'Riordan. "Training to Self-Care: Fitness Tracking, Biopedagogy, and the Healthy Consumer." *Health Sociology Review* 26, no. 1 (2017): 54–68.
Foucault, Michel. *The Birth of the Clinic: An Archaeology of Medical Perception*. New York: Routledge, 1989.
———. *Discipline and Punish: The Birth of the Prison*. New York: Vintage, 1995.
———. *History of Sexuality*. Vol. 3, *The Care of the Self*. New York: Vintage, 1988.
———. *Language, Counter-Memory, Practice: Selected Essays and Interviews*. Edited by Donald F. Bouchard. Ithaca, NY: Cornell University Press, 1977.
———. *The Order of Things: An Archaeology of Human Sciences*. New York: Vintage Books, 1994.
———. *Power/Knowledge: Selected Interviews and Other Writings, 1972–1977*. New York: Pantheon, 1980.
———. *Security, Territory, Population: Lectures at the Collège de France 1977–1978*. New York: Picador, 2008.
Fowler, Geoffrey A. "Wearable Tech Can Spot Coronavirus Symptoms before You Even Realize You're Sick." *Washington Post*, May 28, 2020. www.washingtonpost.com/technology/2020/05/28/wearable-coronavirus-detect/.
Freeman, Jaimie Lee, and Gina Neff. "The Challenge of Repurposed Technologies for Youth: Understanding the Unique Affordances of Digital Self-Tracking for Adolescents." *New Media & Society* 25, no. 11 (2023): 3047–3064.
French, Martin, and Gavin Smith. "'Health' Surveillance: New Modes of Monitoring Bodies, Populations, and Polities." *Critical Public Health* 23, no. 4 (2013): 383–392.
French, Martin, and Gavin J. D. Smith. "Surveillance and Embodiment: Dispositifs of Capture." *Body & Society* 22, no. 2 (2016): 3–27.

Frith, Jordan. "Big Data, Technical Communication, and the Smart City." *Journal of Business and Technical Communication* 31, no. 2 (2017): 168–187.

———. *A Billion Little Pieces: RFID and Infrastructures of Identification*. Cambridge, MA: MIT Press, 2019.

Froth, Marcus, Laura Forlano, Christine Satchell, and Martin Gibbs. *From Social Butterfly to Engaged Citizen: Urban Informatics, Social Media, Ubiquitous Computing, and Mobile Technology to Support Citizen Engagement*. Cambridge, MA: MIT Press, 2011.

Fuller, Joseph, and William Kerr. "The Great Resignation Didn't Start with the Pandemic." *Harvard Business Review*, March 23, 2022. https://hbr.org/2022/03/the-great-resignation-didnt-start-with-the-pandemic.

Fung, Brian. "Is Your Fitbit Wrong? One Woman Argued Hers Was—and Almost Ended up in a Legal No-Man's Land." *Washington Post*, August 2, 2018. www.washingtonpost.com/technology/2018/08/02/is-your-fitbit-wrong-one-woman-argued-it-was-almost-ended-up-legal-no-mans-land/.

Gabriels, Katleen, and Mark Coeckelbergh. "'Technologies of the Self and Other': How Self-Tracking Technologies Also Shape the Other." *Journal of Information, Communication and Ethics in Society* 17, no. 2 (2019): 119–127.

Gaines, Jane. "Political Mimesis." In *Collecting Visible Evidence*, edited by Jane Gaines and Michael Renov, 84–102. Minneapolis: University of Minnesota Press, 1999.

Galloway, Alexander. *Protocol: How Control Exists After Decentralization*. Cambridge, MA: MIT Press, 2004.

Galloway, Alexander R. *The Interface Effect*. Malden, MA: Polity, 2012.

Gardiner, Michael. *Critiques of Everyday Life*. New York: Routledge, 2000.

Gates, Kelly A. *Our Biometric Future: Facial Recognition Technology and the Culture of Surveillance*. New York: New York University Press, 2011.

"Get your Apple Watch with John Hancock Vitality." John Hancock. Accessed September 7, 2019. www.johnhancockinsurance.com/vitality-program/apple-watch.html.

Ghiroli, Brittany, and Rob Biertempfel. "Baseball's New Frontier: How Wearable Technology Is Reshaping the Game." *Athletic*, March 21, 2019. https://theathletic.com/879248/2019/03/21/baseballs-new-frontier-how-wearable-technology-is-reshaping-the-game/.

Gianatasio, David. "George Costanza Needs Google Wallet." *Ad Week*, September 20, 2011. www.adweek.com/creativity/george-costanza-needs-google-wallet-134976/.

Gidaris, Constantine. "Surveillance Capitalism, Datafication, and Unwaged Labour: The Rise of Wearable Fitness devices and Interactive Life Insurance." *Surveillance & Society* 17, nos. 1–2 (2019): 132–138.

Gilbert, Jeremy. "This Conjuncture: For Stuart Hall." *New Formations: A Journal of Culture/Theory/Politics* 96–97 (2019): 38–68.

Gill, Rosalind, and Andy Pratt. "In the Social Factory? Immaterial Labour, Precariousness and Cultural Work." *Theory, Culture, and Society* 25, nos. 7–8 (2008): 1–30.

Gillespie, Tarleton. *Custodians of the Internet: Platforms, Content Moderation, and the Hidden Decisions that Shape Social Media*. New Haven, CT: Yale University Press, 2018.

Gilliom, John, and Torin Monahan. *SuperVision: An Introduction to the Surveillance Society*. Chicago: University of Chicago Press, 2012.

Gilmore, James N. "Alienating and Reorganizing Cultural Goods: Using Lefebvre's Controlled Consumption Model to Theorize Media Industry Change." *International Journal of Communication* 14 (2020): 4474–4493.

———. "Deathlogging: GoPros as Forensic Media in Accidental Sporting Deaths." *Convergence: The International Journal of Research into New Media Technologies* 29, no. 2 (2023): 481–495.

———. "Everywear: The Quantified Self and Wearable Fitness Technologies." *New Media & Society* 18, no. 11 (2016): 2524–2539.

———. "From Ticks and Tocks to Budges and Nudges: The Smartwatch and the Haptics of Informatic Culture." *Television & New Media* 18, no. 3 (2017): 189–202.

———. "Securing the Kids: Geofencing and Child Wearables." *Convergence: The International Journal of Research into New Media Technologies* 26, nos. 5–6 (2020): 1333–1346.

———. "To Affinity and Beyond: Clicking as Communicative Gesture on the Experimentation Platform." *Communication, Culture, and Critique* 13 (2020): 367–383.

Gilmore, James N., and Cassidy Gruber. "Wearable Witnesses: Deathlogging and Framing Wearable Technology Data in 'Fitbit Murders.'" *Mobile Media & Communication* 12, no. 1 (2024): 195–211.

Gitelman, Lisa. *Always Already New: Media, History, and the Data of Culture*. Cambridge, MA: MIT Press, 2008.

———. *"Raw Data" Is an Oxymoron*. Cambridge, MA: MIT Press, 2013.

Gleick, James. *The Information: A History, a Theory, a flood*. New York: Pantheon Books, 2011.

Global Indigenous Data Alliance. Home page. Accessed February 20, 2024. www.gida-global.org.

Goetschel, Max, and Jon M. Phea. "Police Perceptions of Body-Worn Cameras." *American Journal of Criminal Justice* 42 (2017): 698–726.

Goffman, Erving. *The Presentation of Self in Everyday Life*. New York: Anchor Books, 1959.

Goggin, Gerard. "Disability and Haptic Mobile Media." *New Media & Society* 19, no. 10 (2017): 1563–1580.

Goldsmith, Stephen, and Susan Crawford. *The Responsive City: Engaging Communities through Data-Smart Governance*. San Francisco: Jossey-Bass, 2014.

Golumbia, David. "'Communication,' 'Critical.'" *Communication and Critical/Cultural Studies* 10, no. 3 (2013): 248–252.

Goode, Lauren. "Can a Wearable Detect Covid-19 before Symptoms Appear?" *Wired*, April 14, 2020. www.wired.com/story/wearable-covid-19–symptoms-research/.

———. "Fitbit Hit with Class-Action Suit over Inaccurate Heart Rate Monitoring." *Verge*, January 6, 2016. www.theverge.com/2016/1/6/10724270/fitbit-lawsuit-charge-hr-surge-incomplete-heart-rate-tracking.

———. "How Does the Apple Watch Stack up as a health-and-Fitness Tracker?" *Vox*, April 20, 2015. www.vox.com/2015/4/20/11561634/health-and-fitness-on-apple-watch-a-solid-start-with-limitations.

Gordon, Constance, and Kyle Byron. "Sweeping the City: Infrastructure, Informality, and the Politics of Maintenance." *Cultural Studies* 35, nos. 4–5 (2021): 854–875.

"GovDirect Rolls Out Body Worn Cameras to Greenville County Sheriff's." GovDirect, March 10, 2017. www.govdirect.com/blog/govdirect-rolls-out-body-worn-cameras-to-greenville-county-sheriffs.

Graham, Stephen. *Disrupted Cities: When Infrastructure Fails*. New York: Routledge, 2010.

Gramsci, Antonio. *The Prison Notebooks*. Vols. 1–3. New York: Columbia University Press, 2011.

Green, Nicola, and Niles Zurawski. "Surveillance and Ethnography: Researching Surveillance as Everyday Life." *Surveillance & Society* 13, no. 1 (2015): 27–43.

Greenfield, Adam. *Everyware: The Dawning Age of Ubiquitous Computing*. Berkeley, CA: New Riders, 2006.

Greenville County Sheriff's Office-SC. "Greenville County Sheriff's Office CICB-2019–01." Posted March 25, 2019. YouTube video. www.youtube.com/watch?v=fPMhnP83LeY

Greenville Police Department. "Body-Worn Cameras Project." www.greenvillesc.gov/1180/Body-Worn-Cameras-Project.

Gregg, Melissa. *Counterproductive: Time Management in the Knowledge Economy*. Durham, NC: Duke University Press, 2018.

———. *Work's Intimacy*. Malden, MA: Polity, 2011.

Gregory, Sara. "The New Heroes of High School Gym Class: Fitness Trackers." *Roanoke Times*, February 21, 2018. www.roanoke.com/news/education/the-new-heroes-of-high-school-gym-fitness-trackers/article_0ba3f366-384c-5e40-a578-09187e575ac7.html.

Greiner, Jack, and Darren Ford, "Public Access to Police Body Camera Footage—It's Still Not Crystal Clear." *University of Cincinnati Law Review* 86, no. 1 (2018): 139–152.

Grinberg, Daniel. "Tracking Movements: Black Activism, Aerial Surveillance, and Transparency Optics." *Media, Culture & Society* 41, no. 3 (2019): 294–316.

Groening, Stephen. *Cinema beyond Territory: Inflight Entertainment in Global Context*. London: BFI, 2014.

Gross, Daniel J. "As Body Cameras Grow in Popularity, Greenville Police Vehicles Move Away from Dashcams." *Greenville News*, June 26, 2018. www.greenvilleonline.com/story/news/crime/2018/06/26/greenville-police-phase-out-dashcams-rely-body-cameras/714662002/.

———. "As Much Stress as We Put Them under, Officer Training Can't Totally Prepare for Real Crisis." *Greenville News*, September 29, 2019. www.greenvilleonline.com/in-depth/news/local/south-carolina/2019/09/29/sc-police-shootings-training-cant-totally-prepare-officers-for-crisis/3541792002/.

———. "Citizens Review for Deputy Shootings? 'It Won't happen' in Greenville County, Sheriff Says." *Greenville News*, September 29, 2019. www.greenvilleonline.com/in-depth/news/local/south-carolina/2019/09/29/police-accountability-citizens-board-wont-happen-greenville-county/3541936002/.

———. "Greenville County Sheriff's Office Leads SC in Shootings by Law Enforcement." *Greenville News*, September 29, 2019. www.greenvilleonline.com/in-depth/news/local/south-carolina/2019/09/29/police-shootings-in-sc-greenville-county-sheriffs-office-tops-list/779095002/.

———. "Public Won't See Bodycam Video in First Greenville Deputy Shooting under New Program." *Greenville News*, March 13, 2019. www.greenvilleonline.com/story/news/local/south-carolina/2019/03/13/greenville-county-body-cameras/3137811002/.

———. "SC Judges Crack Down on Public Access to Police Body Camera Footage in Greenville." *Greenville News*, June 26, 2019. www.greenvilleonline.com/story/news/local/south-carolina/2019/06/26/greenville-county-sc-judges-crack-down-public-access-police-body-camera/1545756001/.

———. "Secrecy of Police Body Camera Footage in SC Compounds Accountability Issue." *Greenville News*, August 3, 2020. www.greenvilleonline.com/story/news/local/south-carolina/2020/08/03/secrecy-police-body-camera-footage-compounds-accountability-issues/5553837002/.

———. "Some Police Shootings in South Carolina Aren't Captured on Body Camera: Here's Why." *Greenville News*, September 29, 2019. www.greenvilleonline.com/in-depth/news/local/south-carolina/2019/09/29/body-camera-use-sc-police-shootings-is-inconsistent-heres-why/3210213002/.

———. "'We Don't Want to Kill You': Video Shows Tense Moments in Fatal Deputy-Involved Shooting." *Greenville News*, March 25, 2019. www.greenvilleonline.com/story/news/local/south-carolina/2019/03/25/sheriff-deputy-bodycam-shows-tense-moments-fatal-greenville-sc-shooting/3265979002/.

Gross, Daniel J., and Haley Walters. "Greenville, Pickens Body Camera Rule Preventing Lawyers from Sharing Footage Overturned." *Greenville News*, June 27, 2019. www.greenvilleonline.com/story/news/local/south-carolina/2019/06/27/body-camera-order-greenville-pickens-vacated-sc-chief-justice/1586327001/.

Grossberg, Lawrence. "Cultural Studies and Deleuze-Guattari, Part 1: A Polemic on Projects and Possibilities." *Cultural Studies* 28, no. 1 (2014): 1–28.

———. "Cultural Studies in Search of a Method, or Looking for Conjunctural Analysis." *New Formations: A Journal of Culture/Theory/Politics* 96–97 (2019): 38–68.

———. *Cultural Studies in the Future Tense*. Durham, NC: Duke University Press, 2010.

———. *We Gotta Get Out of This Place: Popular Conservatism and Postmodern Culture*. New York: Routledge, 1992,

Guendelsberger, Emily. "I Worked at an Amazon Fulfillment Center; They Treat Workers Like Robots." *Time*, July 18, 2019. https://time.com/5629233/amazon-warehouse-employee-treatment-robots/.

Gunn, Joshua, and Mirko M. Hall. "Stick It in Your Ear: The Psychodynamics of iPod Enjoyment." *Communication and Critical/Cultural Studies* 5, no. 2 (2008): 135–157.

Gunning, Tom. "Chaplin and the Body of Modernity." *Early Popular Visual Culture* 8, no. 3 (2010): 237–245.

Gurdus, Lizzy. "United Healthcare and Fitbit to Pay Users up to $1,500 to Use Devices, Fitbit Co-founder Says." *CNBC*, January 5, 2017. www.cnbc.com/2017/01/05/unitedhealthcare-and-fitbit-to-pay-users-up-to-1500-to-use-devices.html.

Haggerty, Kevin, and Minas Samatras, *Surveillance and Democracy*. New York: Routledge, 2010.

Haggerty, Kevin D., and Richard V. Ericson. "The Surveillant Assemblage." *British Journal of Sociology* 51, no. 4 (2000): 605–622.

Hagood, Mack. "Disability and Biomediation: Tinnitus as Phantom Disability." In *Disability Media Studies*, edited by Elizabeth Ellcessor and Bill Kirkpatrick, 311–329. New York: New York University Press, 2018.

———. *Hush: Media and Sonic Self-Control*. Durham, NC: Duke University Press, 2019.

———. "Quiet Comfort: Noise, Otherness, and the Mobile Production of Personal Space." *American Quarterly* 63, no. 3 (2011): 573–589.

Hale, David. "FSU Ride GPS Technology to Title." *ESPN*, June 22, 2014. www.espn.com/college-football/story/_/id/11121315/florida-state-seminoles-coach-jimbo-fisher-use-gps-technology-win-national-championship.

Haleguoua, Germaine. "The Policy and Export of Ubiquitous Place." In *From Social Butterfly to Engaged Citizen: Urban Informatics, Social Media, Ubiquitous Computing, and Mobile Technology to Support Citizen Engagement*, edited by Marcus Froth, Laura Forlano, Christine Satchell, and Martin Gibbs, 315–334. Cambridge, MA: MIT Press, 2011.

Hall, Rachel, Torin Monahan, and Joshua Reeves. "Editorial: Surveillance and Performance." *Surveillance & Society* 14, no. 2 (2016): 154–167.

Hall, Stuart. *Cultural Studies 1983: A Theoretical History*. Durham, NC: Duke University Press, 2018.

———. "On Postmodernism and Articulation: An Interview with Stuart Hall, Edited by Lawrence Grossberg." In *Stuart Hall: Critical Dialogues in Cultural Studies*, edited by. David Morley and Kuan-Hsing Chen, 131–150. New York: Routledge, 1996.

Hall, Stuart, Chas Critcher, Tony Jefferson, John Clarke, and Brian Roberts. *Policing the Crisis: Mugging, the State, and Law and Order*. New York: Palgrave, 1978.

Hallinan, Blake. "Civilizing Infrastructure." *Cultural Studies* 35, nos. 4–5 (2021): 707–727.

Hallinan, Blake, and James N. Gilmore. "Infrastructural Politics amidst the Coils of Control." *Cultural Studies* 35, nos. 4–5 (2021): 617–640.
Hallinan, Blake, and Ted Striphas. "Recommended for You: The Netflix Prize and the Algorithmic Production of Culture." *New Media & Society* 18, no. 1 (2016): 117–137.
Hartmans, Avery. "The $10,000 Apple Watch Will Stop Getting Major Software Updates from Apple Starting This Fall." *Business Insider*, June 4, 2018. www.business insider.com/apple-watch-gold-edition-will-not-work-with-watchos-5-2018-6.
Hasinoff, Amy Adele. "Where Are You? Location Tracking and the Promise of Child Safety." *Television & New Media* 18, no. 6 (2017): 496–512.
Hassoun, Dan, and James N. Gilmore. "Drowsing: Toward a Concept of Sleepy Screen Engagement." *Communication and Critical/Cultural Studies* 14, no. 2 (2017): 103–119.
Havercroft, Jonathan. "Soul-Blindness, Police Orders, and Black Lives Matter." *Political Theory* 44, no. 6 (2016): 739–763.
Hayles, N. Katherine. "RFID: Human Agency and Meaning in Information-Intensive Environments." *Theory, Culture & Society* 26, nos. 2–3 (2009): 47–72.
Hebdige, Dick. *Subculture: The Meaning of Style*. New York: Routledge, 1979.
Heath, Thomas. "Employee ID Badge Monitors and Listens to You at Work—Except in the Bathroom." *Washington Post*, September 7, 2016. www.washingtonpost.com /news/business/wp/2016/09/07/this-employee-badge-knows-not-only-where -you-are-but-whether-you-are-talking-to-your-co-workers/.
Hecht, Jeffrey. *City of Light: The Story of Fiber Optics*. New York: Oxford University Press, 2004.
Hemmer, Mark. "How Disney Creates Magic with Technology." *OneFire*, October 29, 2015. http://blog.onefire.com/how-disney-creates-magic-with-technology.
"Here One." Home page. Accessed October 16, 2017. http://hereplus.me.
Hern, Alex. "Fitness Tracking App Strava Gives Away Location of Secret US Army Bases." *Guardian*, January 28, 2018. www.theguardian.com/world/2018/jan/28 /fitness-tracking-app-gives-away-location-of-secret-us-army-bases.
Highmore, Ben. *The Everyday Life Reader*. New York: Routledge, 2002.
Hillis, Ken. *Digital Sensations: Space, Identity, and Embodiment in Virtual Reality*. Minneapolis: University of Minnesota Press, 1999.
Holton, Mark. "Walking with Technology: Understanding Mobility-Technology Assemblages." *Mobilities* 14, no. 4 (2019): 435–451.
Hopkins, John, III. "Letter: Shooting Shows Need for Cameras." *Greenville News* April 18, 2015. www.greenvilleonline.com/story/opinion/readers/2015/04/18 /letter-police-shooting-shows-need-police-cameras/25961829/.
"How Does My Tracker Count Steps?" Fitbit. October 28, 2016. https://help.fitbit .com/articles/en_US/Help_article/1143.
Howe, Jonathan E., Wayne L. Black, and Willis A. Jones. "Exercising Power: A Critical Examination of National Collegiate Athletic Association Discourse Related to Name, Image, and Likeness." *Journal of Sport Management* 37, no. 5 (2023): 333–344.

Hsu, Jeremy. "The Strava Heat Map and the End of Secrets." *Wired*, January 29, 2018. www.wired.com/story/strava-heat-map-military-bases-fitness-trackers-privacy/.
Huddleston, Gabriel S., Julie C. Garlen, and Jennifer A. Sandlin. "A New Dimension of Disney Magic: MyMagic+ and Controlled Leisure." In *Disney, Culture, and Curriculum*, edited by Jennifer A. Sandlin and Julie C. Garlen, 220–232. New York: Routledge, 2016.
Humanyze. "Case Study: Technology Company Measures the Impacts of Remote Work to Drive Organizational Health." 2020. http://humanyze.wpengine.com/wp-content/uploads/2020/12/Technology-Company-Measures-Impacts-of-Remote-Work-Case-Study.pdf.
———. "Company." Accessed October 1, 2019. www.humanyze.com/about/Humanyze.
Hunn, Nick. "Hearables—the New Wearables." *Creative Connectivity* (blog), April 3, 2014. www.nickhunn.com/hearables-the-new-wearables/.
Ideo. "A Game-Changing Approach to Sleep for Athletes." Accessed January 10, 2023. www.ideo.com/works/a-game-changing-approach-to-sleep-for-athletes.
Innis, Harold. *Empire and Communications*. Toronto: Dundurn Press, 1950.
"Introducing Fitbit Care." Fitbit Health Solutions. Accessed September 1, 2019. healthsolutions.fitbit.com/fitbit-care/#tools-to-manage-and-measure-your-program.
"Introducing Soundhawk." YouTube video. Accessed September 1, 2019. https://youtu.be/P3ChmlkmKXI.
Jackson, Steven J. "Rethinking Repair." In *Media Technologies: Essays on Communication, Materiality, and Society*, edited by Tarlton Gillespie, P. J. Boczkowski, and Kevin Foot, 221–240. Cambridge, MA: MIT Press, 2014.
Jagoe, Eva-Lynn. "Depersonalized Intimacies: The Cases of Sherry Turkle and Spike Jonze." *ESC* 42, nos. 1–2 (2016): 155–173.
Jenne, Inc. "Panasonic i-PRO Body-Worn Cameras." Posted September 21, 2020. YouTube video. www.youtube.com/watch?v=PZwM_u7fiDI.
Jessop, Alicia, and Thomas A. Baker II. "Big Data Bust: Evaluating the Risks of Tracking NCAA Athletes' Biometric Data." *Texas Review of Entertainment and Sports Law* 20, no. 1 (Fall 2019): 81–112.
Jo, Edward, and Brett A. Dolezal. "Validation of the Fitbit Surge and Charge HR Fitness Trackers." 2016. www.lieffcabraser.com/pdf/Fitbit_Validation_Study.pdf.
John, Nicholas A. *The Age of Sharing*. Malden, MA: Polity, 2016.
———. "File Sharing and the History of Computer: Or, Why File Sharing Is Called 'File Sharing.'" *Critical Studies in Media Communication* 31, no. 3 (2013): 198–211.
"John Hancock Introduces a Whole New Approach to Life Insurance in the U.S. That Rewards Customers for Healthy Living." *PR Newswire*, April 8, 2015. www.prnewswire.com/news-releases/john-hancock-introduces-a-whole-new-approach-to-life-insurance-in-the-us-that-rewards-customers-for-healthy-living-300062461.html.
"John Hancock Leaves Traditional Life Insurance Model Behind to Incentivize Longer, Healthier Lives." John Hancock. September 19, 2018. www.johnhancock.com

/news/insurance/2018/09/john-hancock-leaves-traditional-life-insurance-model-behind-to-incentivize-longer--healthier-lives.html

John Hancock/Vitality. Home page. Accessed September 6, 2019. https://termlife.johnhancockinsurance.com/mvt/vitality-life-quote-combined.

Kalman-Lamb, Nathan, Derek Silva, and Johanna Mellis. "'I Signed My Life to Rich White Guys': Athletes on the Racial Dynamics of College Sports." *Guardian*, March 17, 2021. www.theguardian.com/sport/2021/mar/17/college-sports-racial-dynamics.

Kasson, John F. *Rudeness and Civility: Manners in 19th Century Urban America*. New York: Hill and Wang, 1990.

Kastrenakes, Jacob. "The Biggest Winner from Removing the Headphone Jack Is Apple." *Verge*, September 8, 2016. www.theverge.com/2016/9/8/12839758/apple-is-biggest-winner-from-killing-headphone-jack.

Kelly, Heather. "Amazon's Idea for Employee-Tracking Wearables Raises Concerns." *CNN*, February 2, 2018. https://money.cnn.com/2018/02/02/technology/amazon-employee-tracker/index.html.

Kelly, Mark G. E. "Foucault, Subjectivity, and Technologies of the Self." In *A Companion to Foucault*, edited by Christopher Falzon, Timothy O'Leary, and Jana Sawicki, 510–525. Hoboken, NJ: Blackwell, 2013.

Kennedy, Helen. "Living with Data: Aligning Data Studies and Data Activism through a Focus on Everyday Experiences of Datafication." *Krisis: Journal for Contemporary Philosophy* 38, no. 1 (2018): 18–30.

Kent, Mike, and Katie Ellis. *Disability and New Media*. New York: Routledge, 2010.

Kilgour, Lauren "The Ethics of Aesthetics: Stigma, Information, and the Politics of Electronic Ankle Monitor Design." *Information Society* 36, no. 3 (2020): 131–146.

Kim, Ki Joon, and Dong-Hee Shin. "An Acceptance Model for Smart Watches: Implications for the Adoption of Future Wearable Technology." *Internet Research* 25, no. 4 (2015): 527–541.

Kim, Shin Y., Nicholas P. Deputy, and Cheryl L. Robbins, "Diabetes during Pregnancy: Surveillance, Preconception Care, and Postpartum Care." *Journal of Women's Health* 27, no. 5 (2018): 536–541.

King, Margaret J. "Disneyland and Walt Disney World: Traditional Values in Futuristic Form." *Journal of Popular Culture* 15, no. 1 (1981): 116–140.

Kinnard, Meg. "Officers Concerned about Bill Mandating SC Cops Wear Cameras." *Greenville News*, March 4, 2015. www.greenvilleonline.com/story/news/local/2015/03/04/officers-concerned-bill-mandating-sc-cops-wear-cameras/24379931/.

———. "SC Body Camera Law Spurred by Ferguson, State's Own Shooting." *Greenville News*, August 2, 2015. www.greenvilleonline.com/story/news/local/south-carolina/2015/08/02/sc-body-camera-law-spurred-by-ferguson-states-own-shooting/31023099/.

Kirkpatrick, Bill, and Elizabeth Ellcessor. *Disability Media Studies*. New York: New York University Press, 2017.

Kitchin, Rob. "The Real-Time City? Big Data and Smart Urbanism." *GeoJournal* 79 (2014): 1–14.
Kitchin, Rob, and Martin Dodge. *Code/Space: Software and Everyday Life.* Cambridge, MA: MIT Press, 2014.
Kittler, Friedrich. *Discourse Networks, 1800/1900*. Stanford, CA: Stanford University Press, 1992.
———. *Gramophone, Film, Typewriter*. Stanford, CA: Stanford University Press, 1999.
———. "Thinking Colours and/or Machines." *Theory, Culture, & Society* 23, nos. 7–8 (2006): 39–50.
Kizer, Rebecca. "Fitbit Facing Lawsuit Due to Inaccuracy." *Ball State Daily*, May 30, 2016. www.ballstatedaily.com/article/2016/05/news-fitbit-lawsuit.
Koetsier, John. "Apple Heart Study: What Stanford Medicine Learned from 400,000 Apple Watch Owners." *Forbes*, March 18, 2019. www.forbes.com/sites/johnkoetsier/2019/03/18/apple-heart-study-what-stanford-medicine-learned-from-400000-apple-watch-owners/?sh=48f7c5ee2d20.
Koopman, Colin. *How We Became Our Data: A Genealogy of the Informational Person*. Chicago: University of Chicago Press, 2019.
Korjian, Serge, and C. Michael Gibson. "Digital Technologies and the Democratization of Clinical Research: Social Media, Wearables, and Artificial Intelligence." *Contemporary Clinical Trials* 117 (2022): 106767.
Kornhaber, Donna. "From Posthuman to Postcinema: Crises of Subjecthood and Representation in *Her*." *Cinema Journal* 56, no. 4 (2017): 3–25.
Kristensen, Dorothea Brogard, and Minna Ruckenstein. "Co-evolving with Self-Tracking Technologies." *New Media & Society* 20, no. 10 (2018): 3624–3640.
Kuang, Cliff. "Disney's $1 Billion Bet on a Magical Wristband." *Wired*, March 10, 2015. www.wired.com/2015/03/disney-magicband/.
Kudina, Olya, and Peter-Paul Verbeek. "Ethics from Within: Google Glass, the Collingridge Dilemma, and the Mediated Value of Privacy." *Science, Technology, and Human Values* 44, no. 2 (2019): 291–314.
Laidler, John. "High Tech Is Watching You." *Harvard Gazette*, March 4, 2019. https://news.harvard.edu/gazette/story/2019/03/harvard-professor-says-surveillance-capitalism-is-undermining-democracy/.
Landsbaum, Claire. "Why a Christian University's Freshman Fitbit Requirement Is a Bad Idea." *New York Magazine*, February 4, 2016. http://nymag.com/scienceofus/2016/02/christian-school-requires-fitbits-for-freshmen.html.
Landy, Heather. "See What Happened to Twitter CEO Jack Dorsey's Heart Rate as He Testified to Congress." *Quartz*, September 5, 2018. https://qz.com/work/1380590/twitter-ceo-jack-dorseys-heart-rate-rose-as-he-testified-to-congress.
Lang, Claudia. "Inspecting Mental Health: Depression, Surveillance, and Care in Kerala, South India." *Culture, Medicine and Psychiatry* 43 (2019): 596–612.
Langley, Hugh. "Bragi Exists Wearables as It Sells Dash Business to Mystery Buyer." *Wareable*, April 1, 2019. www.wareable.com/hearables/bragi-exits-consumer-wearables-hearables-7132.

Larkin, Brian. "Degraded Images, Distorted Sounds: Nigerian Video and the Infrastructure of Piracy." *Public Culture* 16, no. 2 (2004): 289–314.

———. *Signal and Noise: Media, Infrastructure, and Urban Culture in Nigeria*. Durham, NC: Duke University Press, 2008.

Latour, Bruno. "'Where Are the Missing Masses? The Sociology of a Few Mundane Artifacts." In *Shaping Technology/Building Society: Studies in Sociotechnical Change*, edited by Wiebe E. Bijker and John Law, 225–228. Cambridge, MA: MIT Press, 1992.

Le Masurier, Guy C., Cara L. Sidman, and Charles B. Corbin. "Accumulating 10,000 Steps: Does This Meet Current Physical Activity Guidelines?" *Research Quarterly for Exercise and Sport* 74, no. 4 (2003): 389–394.

Lebrun, Christopher J. *The Making of Black Lives Matter: A Brief History of an Idea*. New York: Oxford University Press, 2017.

Lee, I-Min, Eric J. Shiroma, Masasmitsu Kamada, David R. Bassett, Charles E. Matthews, and Julie E. Buring. "Association of Step Volume and Intensity with All-Cause Mortality in Older Women." *JAMA Internal Medicine* 179, no. 8 (2019): 1105–1112.

Lefebvre, Henri. *Critique of Everyday Life*. Vol. 2, *Foundations for a Sociology of the Everyday*. New York: Verso, 2004.

———. *Everyday Life in the Modern World*. New York: Routledge, 2017.

———. *Toward an Architecture of Enjoyment*. Minneapolis: University of Minnesota Press, 2014.

Lemire, Joe. "Don't Sleep on the 'U': Miami Football Players Used Own It and Whoop to Get Their Rest and Woke Up Their 2021 season." *Sports Business Journal*, April 29, 2022. www.sportsbusinessjournal.com/Daily/Issues/2022/04/29/Technology/dont-sleep-on-the-u-miami-football-players-set-example-by-using-own-it-and-whoop-to-get-their-rest-and-wake-up-their-2021-season.aspx.

Leskovec, Jute, Anand Rajaraman, and Jeffrey David Ullman. *Mining of Massive Datasets*. Cambridge: Cambridge University Press, 2020.

Leslie, Esther. "This Other Atmosphere: Against Human Resources, Emoji and Devices," *Journal of Visual Culture* 18, no. 1 (2019): 3–29.

Levy, Karen. *Data Driven: Truckers, Technology, and the New Workplace Surveillance*. Princeton, NJ: Princeton University Press, 2022.

Lewis, Michael. *Moneyball: The Art of Winning an Unfair Game*. New York: W. W. Norton, 2004.

Lieberman, Daniel. *Exercised: Why Something We Never Evolved to Do Is Healthy and Rewarding*. New York: Pantheon, 2021.

Lind, Dara. "Obama Wants to Put Body Cameras on 50,000 More Cops." *Vox*, December 1, 2014. www.vox.com/2014/12/1/7314603/body-cameras-police.

Lindsay, Greg. "We Spent Two Weeks Wearing Employee Trackers: Here's What We Learned." *Fast Company*, September 9, 2015. www.fastcompany.com/3051324/we-spent-two-weeks-wearing-employee-trackers-heres-what-we-learned.

Ling, Rich. *Taken-for-Grantedness: The Embedding of Mobile Communication into Society*. Cambridge, MA: MIT Press, 2012.

Lomas, Natasha. "Researchers Spotlight the Lie of 'Anonymous' Data." *Tech Crunch*, July 24, 2019. https://techcrunch.com/2019/07/24/researchers-spotlight-the-lie-of-anonymous-data/.

Lopez, German. "The Failure of Police Body Cameras." *Vox*, July 21, 2017. www.vox.com/policy-and-politics/2017/7/21/15983842/police-body-cameras-failures.

Lowenstein, Adam. *Shocking Representation: Historical Trauma, National Cinema, and the Modern Horror Film*. New York: Columbia University Press, 2008.

Lupton, Deborah. *The Quantified Self*. Malden, MA: Polity, 2016.

Lupton, Deborah, and Ben Williamson. "The Datafied Child: The Dataveillance of Children and Implications for Their Rights." *New Media & Society* 19, no. 5 (2017): 780–794.

Lupton, Deborah, Sarah Pink, and Christine Heyes LaBond. "Digital Traces in Context: Personal Data Contexts, Data Sense, and Self-Tracking Cycling." *International Journal of Communication* 12 (2018): 647–665.

Lyon, David. *The Culture of Surveillance: Watching as a Way of Life*. Malden, MA: Polity, 2018.

———. "Everyday Surveillance: Personal Data and Social Classifications." *Information, Communication, and Society* 5, no. 2 (2002): 242–257.

———. *Surveillance as Social Sorting: Privacy, Risk, and Automated Discrimination*. New York: Routledge, 2003.

Mackenzie, Adrian. *Machine Learners: Archaeology of a Data Practice*. Cambridge, MA: MIT Press, 2017.

———. "The Production of Prediction: What Does Machine Learning Want?" *European Journal of Cultural Studies* 18, nos. 4–5 (2015): 429–455.

Maguire, Donya. "*Affairs of the Phone:* Indiewood, a Bespoke Future, and Virtual Love in Spike Jonze's *Her* (2013)." *Film Matters* (Winter 2016): 49–53.

Mak, Aaron. "What the NBA's $300 COVID-Detecting Rings Can Actually Accomplish." *Slate*, June 22, 2020. https://slate.com/technology/2020/06/nba-coronavirus-oura-ring-orlando.html.

Manjoo, Farhad. "Soundhawk Smart Listening System: A Hearing Helper." *New York Times*, November 19, 2014. www.nytimes.com/2014/11/20/technology/personaltech/soundhawk-smart-listening-system-a-hearing-helper.html.

Mann, Mark. "How Companies Are Using Technology to Make Workers 'Happy' in Their Crappy Jobs." *Vice*, September 5, 2016. www.vice.com/en_us/article/aekn4a/big-data-social-physics-humanyze-tenacity-employment.

Mann, Steve. "New Media and the Power Politics of Sousveillance in a Surveillance-Dominated World." *Surveillance & Society* 11, nos. 1–2 (2013): 18–34.

———. "Sousveillance." WearCam. Accessed July 1, 2024. http://wearcam.org/sousveillance.htm.

———. "Sousveillance: Inventing and Using Wearable Computing Devices for Data Collection in Surveillance Environments." *Surveillance & Society* 1, no. 3 (2003). doi.org/10.24908/ss.v1i33344.

———. "'Sousveillance': Inverse Surveillance in Multimedia Imaging." *Proceedings of the 12th Annual ACM International Conference on Multimedia* (2004): 620–627.

Mannheim, Steve. *Walt Disney and the Quest for Community*. New York: Routledge, 2016.

Mapp, Marqui. "John Hancock's Bargain: Give Us Your Data, You Pay Less in Rates." *CNBC*, April 19, 2015. www.cnbc.com/2015/04/19/john-hancocks-bargain-give-us-more-data-you-pay-less-in-rates.html.

Markortoff, Rebecca. "Study Claims Fitbit Trackers Are 'Highly Inaccurate.'" *CNBC*, May 23, 2016. www.cnbc.com/2016/05/23/study-shows-fitbit-trackers-highly-inaccurate.html.

Martelli, Luca, Katharine Kieslich, and Susi Geiger. "COVID-19 and Techno-solutionism: Responsibilization without Contextualization?" *Critical Public Health* 32, no. 1 (2022): 1–4.

Marwick, Alice. "The Public Domain: Social Surveillance in Everyday Life." *Surveillance and Society* 9, no. 4 (2012): 378–393.

Marx, Karl. *Capital*. Vol. 1, *A Critique of Political Economy*. New York: Penguin, 1992.

Marx, Leo. *The Machine in the Garden: Technology and the Pastoral Ideal in America*. New York: Oxford University Press, 1964.

Mascheroni, Giovanna. "Datafied Childhoods: Contextualizing Datafication in Everyday Life." *Current Sociology* 68, no. 6 (2020): 798–813.

Mattern, Shannon. "Deep Time of Media Infrastructures." In *Signal Traffic: Critical Studies of Media Infrastructures*, edited by Lisa Parks and Nicole Starosielski, 94–114. Urbana: University of Illinois Press, 2015.

Matthews, J. Rosser. *Quantification and the Quest for Medical Certainty*. Princeton, NJ: Princeton University Press, 1995.

Maxmen, Amy, and Jeff Tollefson. "Two Decades of Pandemic War Games Failed to Account for Donald Trump." *Nature* 584, no. 7819 (2020): 26–29.

Mayer, Jonathan, Patrick Mutchler, and John C. Mitchell. "Evaluating the Privacy Properties of Telephone Metadata." *Proceedings of the National Academy of Sciences of the United States of America* 113, no. 20 (2016): 5536–5541.

Mayer-Schönberger, Viktor, and George Cukier. *Big Data: A Revolution That Will Transform How We Live, Work, and Think*. London: John Murray, 2014.

McCann, Jane, and David Bryson. *Smart Clothes and Wearable Technology*. Boca Raton: CRC Press, 2009.

McClusky, Mark. "The Nike Experiment: How the Shoe Giant Unleashed the Power of Personal Metrics." *Wired*, June 22, 2009. www.wired.com/2009/06/lbnp-nike.

McConnell, Kathleen F. "The Profound Sound of Ernest Hemingway's Typist." *Communication and Critical/Cultural Studies* 5 (2008): 325–343.

McCoy, Lauren. "You Have the Right to Tweet, but It Will Be Used against You: Balancing Monitoring and Privacy for Student-Athletes." *Journal of SPORT* 3, no. 2 (2014): 221–245.

McGuire, Dan. "Wearables Conference Explores Fringes of Fashion, Technology." *Tech*, October 21, 1997. LexisNexis Academic.

McMahon, Meghan. "Grappling with the Datafied Self: College Students and Wearable Fitness Trackers," *Columbia University Journal of Politics and Society* 31, no. 1 (2020): 56–88.

Merriam Webster's Online Dictionary. S.v. "Seam." Accessed November 22, 2017. www.merriam-webster.com/dictionary/seam

Michelson, Karl-Jacob. "'Running Is My Boyfriend': Consumers' Relationships with Activities," *Journal of Services Marketing* 31, no. 1 (2017): 24–33.

Milan, Stefania. "Techno-solutionism and the Standard Human in the Making of the COVID-19 Pandemic." *Big Data & Society* 7, no. 2 (2020): 1–7.

Miller, Lisa. "FastPass+: Everything You Need to Know about Walt Disney World's New System." *Huffington Post*, August 2, 2014. www.huffingtonpost.com/2014/05/29/fastpass-plus-disney-world_n_5374335.html.

Millan, Mark. "Apple's Siri Voice Assistant Based in Extensive Research." *CNN*, October 5, 2011. www.cnn.com/2011/10/04/tech/mobile/iphone-siri/index.html.

Miller, Ron. "New Firm Combines Wearables and Data to Improve Decision Making." *Tech Crunch*, February 24, 2015. https://techcrunch.com/2015/02/24/new-firm-combines-wearables-and-data-to-improve-decision-making/.

Mills, Mara. "Hearing Aids and the History of Electronics Miniaturization." *IEEE Annals of the History of Computing* 33, no. 2 (2011): 24–44.

Milne, Esther. *Email and the Everyday: Stories of Disclosure, Trust, and Digital Labor*. Cambridge, MA: MIT Press, 2021.

MIT Technology Review Insights. "Technology for Workplaces That Work: Humanyze's Ben Waber." *MIT Technology Review*, January 24, 2019. www.technologyreview.com/2019/01/24/137732/technology-for-workplaces-that-work-humanyzes-ben-waber/.

Mitchell, David T., and Sharon L. Snyder. *The Biopolitics of Disability: Neoliberalism, Ablenationalism, and Peripheral Embodiments*. Ann Arbor: University of Michigan Press, 2015.

Mitchell, W. J. T. *Cloning Terror: The War of Images, 9/11 to the Present*. Chicago: University of Chicago Press, 2011.

Mitrasinovic, Miodrag. *Total Landscape, Theme Parks, Public Space*. New York: Routledge, 2006.

Monaghan, Lee F., Rachel Colls, and Bethan Evans. "Obesity Discourse and Fat Politics: Research, Critique, and Interventions." *Critical Public Health* 23, no. 3 (2013): 249–262.

Monahan, Torin. "Reckoning with COVID, Racial Violence, and the Perilous Pursuit of Transparency." *Surveillance & Society 19*, no. 1 (2021): 1–10.

"Monitor Your Heart Rate with Apple Watch." Apple. Accessed May 20, 2023. https://support.apple.com/en-us/HT204666.

Moore, Phoebe V. *The Quantified Self in Precarity: Work, Technology, and What Counts*. New York: Routledge, 2018.

Morozov, Evgeny. *To Save Everything, Click Here: The Folly of Technological Solutionism*. New York: Public Affairs, 2013.

Morris, Jeremy Wade, and Sarah Murray. *Appified: Culture in the Age of Apps*. Ann Arbor: University of Michigan Press, 2018.

Mossberg, Walter S., and Katherine Boehret. "On the Run with the iPod/Nike Fitness Device." *Wall Street Journal*, July 16, 2006. www.wsj.com/articles/SB115326608907010478.

Motsinger, Carol, and Daniel J. Gross. "SC Fails to Abide by Spirit of Its Own Law as Police Videos Spur Demands for Justice." *Greenville News*, August 3, 2020. www.greenvilleonline.com/story/news/local/south-carolina/2020/08/03/body-cameras-funding-oversight-lacking-5-years-after-new-law-sc/5341127002/.

Mull, Amanda. "What 10,000 Steps Will Really Get You." *Atlantic*, May 31, 2019. www.theatlantic.com/health/archive/2019/05/10000–steps-rule/590785/.

Mullin, Emily. "Why Siri Won't Listen to Millions of People with Disabilities." *Scientific American*, May 27, 2016. www.scientificamerican.com/article/why-siri-won-t-listen-to-millions-of-people-with-disabilities/.

Mumford, Lewis. *Technics and Civilization*. Chicago: University of Chicago Press, 1934.

Nafus, Dawn. "The Domestication of Data: Why Embracing Digital Data Means Embracing Bigger Questions." *Ethnographic Praxis in Industry Conference Proceedings*, no. 1 (2016): 387–399.

———. *Quantified: Biosensing Technologies in Everyday Life*. Cambridge, MA: MIT Press, 2016.

Nagy, Peter, and Gina Neff. "Imagined Affordance: Reconstructing a Keyword for Communication Theory." *Social Media + Society* (July–December 2015): 1–9.

Natale, Simone. *Deceitful Media: Artificial Intelligence and Social Life after the Turing Test*. New York: Oxford University Press, 2021.

Neff, Gina, and Dawn Nafus. *Self-Tracking*. Cambridge, MA: MIT Press, 2016.

Negroponte, Nicholas. *Being Digital*. London: Hodder and Stoughton, 1995.

Nelkin, Dorothy, and Lori Andrews. "DNA Identification and Surveillance Creep." *Sociology of Health and Illness* 21, no. 5 (1999): 689–706.

Nelson, Alondra. "The Longue Duree of Black Lives Matter." *American Journal of Public Health* 106, no. 10 (2016): 1734–1737.

New, Jake. "Auburn Has a Private Security Firm Enforcing Players' Nightly Curfews." *Deadspin* (blog). November 8, 2012. https://deadspin.com/auburn-has-a-private-security-firm-enforcing-players-ni-5958852.

———. "Class Checkers." *Inside Higher Ed*, June 23, 2015. www.insidehighered.com/news/2015/06/24/attendance-monitoring-programs-common-college-athletics.

"New Student Orientation Schedule." Oral Roberts University. 2018. www.oru.edu/oru-experience/first-year-experience/orientation/schedule.php.

Newcomb, Horace, and Paul M. Hirsch. "Television as a Cultural Forum: Implications for Research." *Quarterly Review of Film and Video* 8, no. 3 (1983): 45–55.

Newman, Lily Hay. "Police Bodycams Can Be Hacked to Doctor Footage." *Wired*, August 11, 2018. www.wired.com/story/police-body-camera-vulnerabilities.

Nichols, Bill. *Representing Reality: Issues and Concepts in Documentary*. Bloomington: Indiana University Press, 1991.

Nicholls, Rochelle, Glenn Fleisig, Bruce Elliott, Stephen Lyman, and Edmund Osinski. "Baseball: Accuracy of Qualitative Analysis for Assessment of Skilled Baseball Pitching Technique." *Sports Biomechanics* 2, no. 2 (2003): 213–226.

"Nike iPod Tune Your Run OK Go." Posted December 1, 2006. YouTube video. www.youtube.com/watch?v=HEs8NIRRyYc.

Noyes, Katherine. "Humanyze's 'People Analytics' Wants to Transform Your Workplace." *Computer World*, November 20, 2015. www.computerworld.com/article/3006631/startup-humanyzes-people-analytics-wants-to-transform-your-workplace.html.

Nusca, Andrew. "Say Command: How Speech Recognition Will Change the World." *ZDNet*, November 2, 2011. www.zdnet.com/article/say-command-how-speech-recognition-will-change-the-world.

Obar, Jonathan A., and Anne Oeldorf-Hirsch. "The Biggest Lie on the Internet: Ignoring the Privacy Policies and Terms of Service Policies of Social Networking Sites." *Information, Communication, & Society* 23, no. 1 (2020): 128–147.

Oh, Jeeyun, and Hyunjin Kant. "User Engagement with Smart Wearables: Four Defining Factors and a Process Model." *Mobile Media & Communication* 9, no. 2 (2021): 314–335.

O'Neill, Christopher. "Taylorism, the European Science of Work, and the Quantified Self at Work." *Science, Technology, and Human Values* 42, no. 4 (2017): 600–621.

Ong, Thuy. "Amazon Patents Wristbands That Track Warehouse Employees' Hands in Real Time." *Verge*, February 1, 2018. www.theverge.com/2018/2/1/16958918/amazon-patents-trackable-wristband-warehouse-employees.

Online Etymology Dictionary. S,v. "Band." Accessed October 20, 2017. www.etymonline.com/index.php?term=bind&allowed_in_frame=0.

Oremus, Will. "These Aren't Wireless Headphones." *Slate*, September 8, 2016. www.slate.com/articles/technology/future_tense/2016/09/apple_s_airpods_aren_t_just_wireless_earbuds_they_re_the_future_of_computing.html.

Oudart, Jean-Pierre. "Notes on Suture." *Screen* 18, no. 4 (1977–78): 35–47.

Oura. "The Oura Difference." Oura Ring. 2020. https://ouraring.com/the-Oura-difference.

———. "Readiness: Your Complete Guide." Oura Ring. 2020. https://ouraring.com/readiness-score.

———. "Terms of Use." Oura Ring. 2018. https://ouraring.com/terms-and-conditions.

Oxford English Dictionary. S.v. "Apparatus." Accessed November 5, 2022. https://en.oxforddictionaries.com/definition/apparatus.

Pagliarella, Chris. "Police Body-Worn Camera Footage: A Question of Access." *Yale Policy & Law Review* 34 (2016): 532–543.

Palmer, Darren. "The Mythical Properties of Police Body-Worn Cameras: A Solution in the Search of a Problem." *Surveillance & Society* 14, no. 1 (2016): 138–144.

Panasonic Business Solutions. "NY1 Looks at the Panasonic Arbitrator BWC." Posted March 8, 2017. YouTube video. www.youtube.com/watch?v=tm8Pn4c3YZ4.

———. "Panasonic Arbitrator BWC." Posted March 9, 2017. YouTube video. www.youtube.com/watch?v=NVa7SJaRkLI.

Pariser, Eli. *The Filter Bubble: What the Internet Is Hiding from You*. New York: Penguin Books, 2012.

Parisi, David. "Game Interfaces as Disabling Infrastructures." *Analog Game Studies* 4, no. 3 (2017). http://analoggamestudies.org/2017/05/compatibility-test-videogames-as-disabling-infrastructures/.

Parks, Lisa. *Cultures in Orbit: Satellites and the Televisual*. Durham, NC: Duke University Press, 2005.

Parks, Lisa, and Nicole Starosielski. *Signal Traffic: Critical Studies of Media Infrastructures*. Urbana: University of Illinois Press, 2015.

Pasternak, Harley. "The Power of Taking 10,000 Steps (or More!) and How to Get There." *Fitbit* (blog),. March 23, 2018. https://blog.fitbit.com/walking-10000-steps-a-day.

Payne, Robert. "Frictionless Sharing and Digital Promiscuity." *Communication and Critical/Cultural Studies* 11, no. 2 (2014): 85–102.

Peirce, Charles. *Collected Writings*. Edited by Charles Hartshorne, Paul Weiss, and Arthur Banks. Cambridge, MA: Harvard University Press, 1965.

Penley, Constance, and Andrew Ross. *Technoculture*. Minneapolis: University of Minnesota Press, 1991.

Pentland, Alex. *Social Physics: How Social Networks Can Make Us Smarter*. New York: Penguin Books, 2015.

Perez, Marco V., Kenneth W. Mahaffey, Haley Heflin, John S Rumsfeld, Ariadna Garcia, Todd Ferris, Vidhya Balasubramanian, Andrea M. Russo, Amol Rajmane, Lauren Cheung, Grace Hung, Justin Lee, Peter Kowey, Nisha Talati, Divya Nag, Santosh E. Gummidipundi, Alexis Beatty, Mellanie True Hills, Sumbul Desai, Christopher B. Granger, Manisha Desai, and Mintu P. Turakhia. "Large-Scale Assessment of a Smartwatch to Identify Atrial Fibrillation." *New England Journal of Medicine* 381 (2019): 1909–1917.

Petchesky, Ben. "Auburn Has a Private Security Firm Enforcing Players' Nightly Curfews." *Deadspin* (blog). November 8, 2012. https://deadspin.com/auburn-has-a-private-security-firm-enforcing-players-ni-5958852.

Peters, John Durham. *The Marvelous Clouds: Toward a Philosophy of Elemental Media*. Chicago: University of Chicago Press, 2015.

———. "'You Mean My Whole Fallacy Is Wwrong': On Technological Determinism." *Representations* 140 (2017): 10–26.

Petroski, Henry. *The Road Taken: The History and Future of America's Infrastructure*. New York: Bloomsbury, 2016.

Picard, Rosalind. *Affective Computing*. Cambridge, MA: MIT Press, 2000.

Pickard, Victor. "Being Critical: Contesting Power within the Misinformation Society." *Communication and Critical/Cultural Studies* 10, no. 3 (2013): 306–311.

Pickering, Michael. *History, Experience, and Cultural Studies*. New York: St. Martin's Press, 1997.

Pickman, Ben. "The Story behind the Ring That Is Key to the NBA's Restart." *Sports Illustrated*, July 1, 2020. www.si.com/nba/2020/07/01/oura-ring-nba-restart-orlando-coronavirus.

Pierce, David. "Bragi Dash Puts a New Kind of Computer in Your Ears." *Wired*, January 11, 2016. www.wired.com/2016/01/bragi-dash/.

———. "Inside the Downfall of Doppler Labs." *Wired*, November 1, 2017. www.wired.com/story/inside-the-downfall-of-doppler-labs/.

Pike, David L. "The Walt Disney World Underground." *Space and Culture* 8, no. 1 (2005): 47–65.

Pink, Sarah, and Vaike Fors. "Self-Tracking and Mobile Media: New Digital Materialities." *Mobile Media & Communication* 5, no. 3 (2017): 219–238.

Plotnick, Rachel. "Touch of a Button: Long-Distance Transmission, Communication, and Control at World's Fairs." *Critical Studies in Media Communication* 30, no. 1 (2013): 52–68.

Pogue, David. "Make Technology—and the World—Frictionless." *Scientific American*, April 1, 2012. www.scientificamerican.com/article/technologys-friction-problem/.

Polar. "Making Physical Education Measurably More Fun." Accessed June 20, 2018. www.polar.com/en/business/education/.

———. "Privacy Notice." Accessed June 20, 2018. www.polar.com/us-en/legal/privacy-notice#toc25.

"Police Body Cameras: Money for Nothing?" *Fox News*, October 21, 2017. www.foxnews.com/politics/police-body-cameras-money-for-nothing.

Poster, Winifred. "Emotion Detectors, Answering Machines and E-unions: Multi-surveillance in the Global Interactive Services Industry." *American Behavioral Scientist* 55, no. 7 (2011): 868–901.

Powers, Devon. *On Trend: The Business of Forecasting the Future*. Urbana-Champaign: University of Illinois Press, 2019.

Prasopolou, Eloisa. "A Half-Moon on My Skin: A Memoir on Life with an Activity Tracker." *European Journal of Information Systems* 26, no. 3 (2017): 287–297.

Previte, Samantha. "NBA to Use 'Smart Rings,' Big Data to Fight Coronavirus in Disney Bubble." *New York Post*, June 19, 2020. https://nypost.com/2020/06/19/nba-to-use-smart-rings-to-detect-coronavirus-within-bubble/.

"Product Details." Here One. Accessed October 26, 2017. https://hereplus.me/pages/product-details.

Pyysiäinen, Jarkko, Darren Halpin, and Andrew Guilfoyle. "Neoliberal Governance and 'Responsibilization' of Agents: Reassessing the Mechanisms of Responsibility-Shift in Neoliberal Discursive Environments." *Distinktion: Journal of Social Theory* 18, no. 2 (2017): 215–235.

Reeves, Joshua. "Automatic for the People: The Automation of Communicative Labor." *Communication and Critical/Cultural Studies* 13, no. 2 (2016): 150–165.

"Regulatory Requirements for Hearing Aid Devices and Personal Sound Amplification Products—Draft Guidance for Industry and Food and Drug Administration

Staff." US Food and Drug Administration. November 7, 2013. www.fda.gov/MedicalDevices/ucm373461.htm.

Reichert, Corinne. "NBA Players Could Wear Smart Ring to track COVID-19 Symptoms as Season Resumes." *CNET*, June 22, 2020. www.cnet.com/tech/mobile/nba-players-could-wear-a-smart-ring-to-track-covid-19-symptoms-as-season-resumes-at-disney-world/.

Reynolds, C. J. "Mischievous Infrastructure: Tactical Secrecy through Infrastructural Friction in Police Video Systems." *Cultural Studies* 35, nos. 4–5 (2021): 996–1019.

Rhodes, Margaret. "A Sleek New Hearing Aid That Solves a Nagging Problem." *Wired*, June 24, 2014. www.wired.com/2014/06/a-sleek-new-hearing-aid-that-solves-a-nagging-problem.

Ripley, Amanda. "A Big Test of Police Body Cameras Defies Expectations." *New York Times*, October 20, 2017. www.nytimes.com/2017/10/20/upshot/a-big-test-of-police-body-cameras-defies-expectations.html.

Ristić, Dušan, and Dušan Marinkovič, "Lifelogging: Digital Technologies of the Self as Practices of Contemporary Biopolitics," *Siologija* 4 (2019): 535–549.

Robb, Drew. "Building the Global Heatmap." Medium. November 1, 2017. https://medium.com/strava-engineering/the-global-heatmap-now-6x-hotter-23fc01d301de.

Roberts-Lewis, Sarah F., Claire M. White, Mark Ashworth, and Michael R. Rose. "Validity of Fitbit Activity Monitoring for Adults with Progressive Muscle Diseases." *Disability and Rehabilitation* 44, no. 24 (2022): 7543–7553.

Robertson, Adi. "A Wisconsin Company Will Let Employees Use Microchip Implants to Buy Snacks and Open Doors." *Verge*, July 24, 2017. www.theverge.com/2017/7/24/16019530/three-sqaure-market-implant-office-keycard-biohacking-wisconsin.

Rockefeller Neuroscience Institute. "Understanding the Spread: Protecting Our Health and Economy." 2020. http://wvumedicine.org/RNI/COVID19.

Rose, Nikolas. *Powers of Freedom: Reframing Political Thought*. Cambridge: Cambridge University Press, 1999.

Ross, Andrew. *The Celebration Chronicles: Life, Liberty, and the Pursuit of Property Value in Disney's New Town*. New York: Ballantine Books, 1999.

Rothman, William. "Against 'The System of the Suture.'" *Film Quarterly* 29, no. 1 (1975): 45–50.

Ruckenstein, Minna. "Visualized and Interacted Life: Personal Analytics and Engagements with Data Doubles." *Societies* 4, no. 1 (2014): 68–84.

Ruckenstein, Minna, and Mika Pantzar. "Beyond the Quantified Self: Thematic Exploration of a Dataistic Paradigm." *New Media & Society* 19, no. 3 (2017): 401–418.

Ryan, Susan Elizabeth. *Garments of Paradise: Wearable Discourse in the Digital Age*. Cambridge, MA: MIT Press, 2014.

Sabin, Rainer. "Inside the Technology Giving Alabama a Competitive Edge." Alabama Football. July 2, 2017. www.al.com/alabamafootball/2017/07/inside_the_technology_giving_a.html.

Sadowski, Jathan. "When Data Is Capital: Datafication, Accumulation, and Extraction." *Big Data & Society* 6, no. 1 (2019): 1–12. doi.org/2053951718820549.

Salamone, Virginia A., and Frank A. Salamone. "Images of Main Street: Disney World and the American Adventure." *Journal of American Culture* 22, no. 1 (1999): 85–93.

Sanderson, Jimmy, Melinda Weathers, Katherine Snedaker, and Kelly Gramlich. "'I Was Able to Still Do My Job on the Field and Keep Playing': An Investigation of Female and Male Athletes' Experiences with (Not) Reporting Concussions." *Communication and Sport* 5, no. 3 (2017): 267–287.

Sanfilippo, Madelyn Rose, and Yan Shvartzshnaider. "Data and Privacy in a Quasi-Public Space: Disney World as a Smart City." In *Diversity, Divergence, Dialogue—16th International Conference, iConference 2021, Proceedings*, edited by Katharina Toeppe, Hui Yan, and Samuel Kahi Chu, 235–250. Cham, Switzerland: Springer, 2021.

Saval, Nikal. *Cubed: A Secret History of the Workplace*. New York: Doubleday, 2014.

Scheufele, Dietram A. "Framing as a Theory of Media Effects." *Journal of Communication* 49, no. 1 (1999): 103–122.

Schmidt, Michael S., and Matt Apuzzo. "South Carolina Officer Is Charged with Murder of Walter Scott." *New York Times*, April 8, 2015. www.nytimes.com/2015/04/08/us/south-carolina-officer-is-charged-with-murder-in-black-mans-death.html.

Schofield, Jack. "From Man to Borg—Is This the Future?" *Guardian*, August 2, 2001. www.theguardian.com/technology/2001/aug/02/onlinesupplement.gadgets.

Schüll, Natasha Dow. "Data for Life: Wearable Technology and the Design of Self-Care." *BioSocieties* 11, no. 3 (2016): 317–333.

Schultheis, Emily. "Hillary Clinton Calls for Body Cameras for All Police Officers Nationwide." *Atlantic*, April 29, 2015. www.theatlantic.com/politics/archive/2015/04/hillary-clinton-calls-for-body-cameras-for-all-police-officers-nationwide/457815/.

Shapiro, Aaron. *Design, Control Predict: Logistical Governance in the Smart City*. Minneapolis: University of Minnesota Press, 2020.

Sharma, Sarah. *In the Meantime: Temporality and Cultural Politics*. Durham, NC: Duke University Press, 2014.

Shaw, Adrienne. "Encoding and Decoding Affordances: Stuart Hall and Interactive Media Technologies." *Media, Culture, and Society* 39, no. 4 (2017): 592–602.

Sheng, Ellen. "Employee Privacy in the US Is at Stake as Corporate Surveillance Technology Monitors Workers' Every Move." *CNBC*, April 15, 2019. www.cnbc.com/2019/04/15/employee-privacy-is-at-stake-as-surveillance-tech-monitors-workers.html.

Shepard, Mark. *Sentient City: Ubiquitous Computing, Architecture, and the Future of Urban Space*. Cambridge, MA: MIT Press, 2011.

Shih, Patti, Kathleen Prkopovich, Chris Degeling, Jacqueline Street, and Stacy M. Carter. "Direct-to-Consumer Detection of Atrial Fibrillation in a Smartwatch

Electrocardiogram: Medical Overuse, Medical Inaction and the Experience of Consumers," *Social Science & Medicine* 303 (2022): 114954.

Simon, Matt. "Inside the Amazon Warehouse Where Humans and Machines Become One." *Wired*, June 5, 2019. www.wired.com/story/amazon-warehouse-robots/.

Slack, Jennifer Daryl. "The Theory and Method of Articulation in Cultural Studies." In *Stuart Hall: Critical Dialogues in Cultural Studies*, edited by David Morley and Kuan-Hsing Chen, 112–127. New York: Routledge, 1992.

Slack, Jennifer Daryl, and J. Macgregor Wise. *Culture and Technology: A Primer*. 2nd ed. New York: Peter Lang, 2015.

Smarr, Benjamin L., Kirstin Aschbacher, Sarah M. Fischer, Anoushka Chowdhary, Stephan Dilchert, Karena Puldon, Adam Rao, Frederick M. Hecht, and Ashley E. Mason. "Feasibility of Continuous Fever Monitoring Using Wearable Devices." *Scientific Reports* 10 (2020): 21640.

Smiley, Lauren. "A Brutal Murder, a Wearable Witness, and an Unlikely Suspect." *Wired*, September 17, 2019. www.wired.com/story/telltale-heart-fitbit-murder/.

Smith, Chris. "A Day out with Disney's MagicBand 2." *Wareable*, April 11, 2017. www.wareable.com/wearable-tech/disney-magicband-2–review.

Smith, Neil. "Homeless/Global: Scaling Places." In *Mapping the Futures: Local Cultures, Global Change*, edited by John Bird, Barry Curtis, Tim Putnam, George Robertson, and Lisa Tickner, 87–119. New York: Routledge, 1993.

Smith, Tim. "Policy Body Cam Rules Detail Who Should Wear Them and When." *Greenville News*, December 4, 2015. www.greenvilleonline.com/story/news/2015/12/04/body-camera-guidelines-approved-sc-law-enforcement/76777168/.

Snow, C. P. *The Two Cultures*. Cambridge: Cambridge University Press, 2014.

Snyder, Scott, and Alex Castrounis. "How to Turn 'Data Exhaust' into a Competitive Edge." *Knowledge at Wharton*, March 1, 2018. https://knowledge.wharton.upenn.edu/article/turn-iot-data-exhaust-next-competitive-advantage.

Solove, Daniel J. "'I've Got Nothing to Hide' and Other Misunderstandings of Privacy." *San Diego Law Review* 44 (2007): 745–772.

Song, Victoria. "Only Athletes Should Give a Whoop about Whoop." *Verge*, March 2, 2022. www.theverge.com/22957195/whoop-review-fitness-tracker-wearables.

Staggs, Tom. "Taking the Disney Guest Experience to the Next Level." *Walt Disney World Report* (blog), January 7, 2013. https://disneyparks.disney.go.com/blog/2013/01/taking-the-disney-guest-experience-to-the-next-level/.

Stanford Medicine. "Through Apple Heart Study, Stanford Medicine Researchers Show Wearable Technology Can Help Detect Atrial Fibrillation." Stanford Medicine News Center. November 13, 2019. https://med.stanford.edu/news/all-news/2019/11/through-apple-heart-study--stanford-medicine-researchers-show-we.html.

"Stanford Medicine Announces Results of Unprecedented Apple Heart Study." Apple. March 16, 2019. www.apple.com/newsroom/2019/03/stanford-medicine-announces-results-of-unprecedented-apple-heart-study.

Stark, Luke. "Algorithmic Psychometrics and the Scalable Subject." *Social Studies of Science* 48, no. 2 (2018): 204–231.

Starosielski, Nicole. *The Undersea Network*. Durham, NC: Duke University Press, 2015.
Sterne, Jonathan. *The Audible Past: Cultural Origins of Sound Reproduction*. Durham, NC: Duke University Press, 2003.
———. *Diminished Faculties: A Political Phenomenology of Impairment*. Durham, NC: Duke University Press, 2022.
———. *MP3: The Meaning of a Format*. Durham, NC: Duke University Press, 2012.
Stoddard, Eric. "A Surveillance of Care: Evaluating Surveillance Ethically." In *Routledge Handbook of Surveillance Studies*, edited by Kirstie Ball, Kevin Haggerty, and David Lyon, 369–376. New York: Routledge.
Stone, Jeff. "Not All Oral Roberts Students Need to Wear Fitbits, and They're Not Tracked through Campus." *International Business Times*, February 3, 2016. www.ibtimes.com/not-all-oral-roberts-students-need-wear-fitbits-theyre-not-tracked-through-campus-2291808.
Stone, Kaitlyn. "Enter the World of Yesterday, Tomorrow, and Fantasy: Walt Disney World's Creation and Its Implication on Privacy Rights under the MagicBand System." *Journal of High Technology Law* 18, no. 1 (2017): 198–238.
Striphas, Ted. "Algorithmic Culture." *European Journal of Cultural Studies* 18, nos. 4–5 (2015): 395–412.
———. *Algorithmic Culture before the Internet*. New York: Columbia University Press, 2023.
———. "Caring for Cultural Studies." *Cultural Studies* 33, no.1 (2019): 1–18.
———. "Keyword: Critical." *Communication and Critical/Cultural Studies* 10, no. 3 (2013): 324–328.
———. "Known-Unknowns: Matthew Arnold, F. R. Leavis, and the Government of Culture." *Cultural Studies* 31, no. 1 (2017): 143–163.
———. *The Late Age of Print: Everyday Book Culture from Consumerism to Control*. New York: Columbia University Press, 2009.
———. "The Visible College." *International Journal of Communication* 5 (2011): 1744–1751.
Sullivan, Paul. "Life Insurance Offering More Incentive to Live Longer." *New York Times*, September 19, 2018. www.nytimes.com/2018/09/19/your-money/john-hancock-vitality-life-insurance.html.
Swan, Melanie. "The Quantified Self: Fundamental Disruption in Big Data Science and Biological Discovery." *Big Data* 1, no. 2 (2013): 85–99.
Swartz, Lana. *New Money: How Payment became Social Media*. New Haven, CT: Yale University Press, 2020.
Taylor, Brian. "Hearables." *Audiology Today* 27, no. 6 (2015): 22–30.
Taylor, Charles. *Modern Social Imaginaries*. Durham, NC: Duke University Press, 2004.
Taylor, Emmaline. "Lights, Camera, Redaction . . . Police Body-Worn Cameras: Autonomy, Discretion and Accountability." *Surveillance & Society* 14, no. 1 (2016): 128–132.

Taylor, Linnet. "There Is an App for That: Technological Solutionism as COVID-19 Policy in the Global North." In *The New Common: How the COVID-19 Pandemic Is Transforming Society*, edited by Emile Aarts, Hein Fleuren, Margriet Sitskoorn, and Ton Wilthagen, 209–216. Cham, Switzerland: Springer, 2021.

"Technology in the Oura Ring." Oura. Accessed July 28, 2020. https://ouraring.com/blog/ring-technology/

Thacker, Eugene. "Biomedia." In *Critical Terms for Media Studies*, edited by W. J. T. Mitchell and Miriam B. N. Hansen, 117–130. Chicago: University of Chicago Press, 2010.

Thaler, Richard H., and Cass R. Sunstein. *Nudge: Improving Decisions about Health, Wealth, and Happiness*. New York: Penguin Books, 2009.

Thompson, E. P. "Time, Work-Discipline, and Industrial Capitalism." *Past & Present* 38 (1967): 56–97.

"Tips for Getting Your Steps In." Centers for Disease Control and Prevention. May 9, 2016. www.cdc.gov/features/getting-your-steps-in/index.html.

Toonders, Jonis. "Data Is the New Oil of the digital economy." *Wired*, July 2014. www.wired.com/insights/2014/07/data-new-oil-digital-economy/.

Topol, Eric. *The Patient Will See You Now: The Future of Medicine is in Your Hands*. New York: Basic Books, 2015.

Townsend, Anthony. *Smart Cities: Big Data, Civic Hackers, and the Quest for a New Utopia*. New York: Norton, 2013.

Tracy, Marc. "Technology Used to Track Players' Steps Now Charts Their Sleep, Too." *New York Times*, September 22, 2017. www.nytimes.com/2017/09/22/sports/ncaafootball/clemson-alabama-wearable-technology.html.

———. "With Wearable Tech Deals, New Player Data Is Up for Grabs." *New York Times*, September 11, 2016, www.nytimes.com/2016/09/11/sports/ncaafootball/wearable-technology-nike-privacy-college-football.html

Trevorrow, Philippa. "Technology Running the World: The Nike+iPod Kit and Levels of Physical Activity." *Society and Leisure* 35, no. 1 (2012): 131–154.

Troyan, Mary. "Scott Wants $500 Million for Police Body Cameras." *Greenville News*, July 18, 2015. www.greenvilleonline.com/story/news/local/2015/07/28/tim-scott-police-body-cameras/30795815/.

Tu, Rungting, Peishan Hsieh, and Wenting Feng, "Walking for Fun or for 'Likes'? The Impacts of Different Gamification Orientations of Fitness Apps on Consumers' Physical Activities," *Sport Management Review* 22, no. 5 (2019): 682–693.

Tufekci, Zeynep. "Engineering the Public: Big data, Surveillance and Computational Publics." *First Monday* 19, no. 7 (2014). https://journals.uic.edu/ojs/index.php/fm/article/view/4901.

Turkle, Sherry. *Alone Together: Why We Expect More from Technology and Less from Each Other*. New York: Basic Books, 2012.

———. "Always-on/Always-on-You: The Tethered Self." In *Handbook of Mobile Communication Studies*, edited by James E. Katz, 121–138. Cambridge, MA: MIT Press, 2008.

Turow, Joseph. *The Daily You: How the New Advertising Industry Is Defining Your Identity and Your Worth*. New Haven, CT: Yale University Press, 2013.
Vaidhyanathan, Siva. *The Googlization of Everything (and Why We Should Worry)*. Berkeley: University of California Press, 2011.
van Dijck, José. "Datafication, Dataism, and Dataveillance: Big Data between Scientific Paradigm and Ideology." *Surveillance and Society* 12, no. 2 (2014): 197–208.
Veblen, Thorstein. *Theory of the Leisure Class*. New York: Dover Print, 1994. First published 1899.
Verge, The. "Bragi Dash Wireless Earbuds Review." Posted March 18, 2016. YouTube video. www.youtube.com/watch?v=BTLhQ11snU8.
Waber, Ben. "Decoding Workforce Productivity." World Economic Forum. Posted August 8, 2016. YouTube video. www.youtube.com/watch?v=i-F7Cd_W4Uc.
———. "Don't Let COVID-19 Compromise Your Organizational Health." Humanyze. March 10, 2021. https://humanyze.com/dont-let-covid-19-compromise-your-organizational-health/.
———. "How Companies Can Maintain Organizational Health amid COVID-19." *TLNT*, May 27, 2020. www.tlnt.com/articles/how-companies-can-maintain-organizational-health-amid-covid-19.
———. *People Analytics: How Social Sensing Technology Will Transform Business and What It Tells Us about the Future of Work*. Upper Saddle River, NJ: FT Press, 2013.
———. *People Analytics*. Upper Saddle River, NJ: FT Press, 2013. E-book on O'Reilly learning platform. https://learning.oreilly.com/library/view/people-analytics-how/9780133158342/toc.html.
———. "Using Analytics to Measure Interactions in the Workplace." re:Work with Google. Posted November 10, 2014. YouTube video. www.youtube.com/watch?v=XojhyhoRI7I.
———. "Work-Life Balance in the Time of Coronavirus." *Humanyze* (blog). Accessed November 10, 2022. https://humanyze.com/blog-work-life-balance-during-coronavirus/.
Wallet, Matthew P., Susan R. Gomersall, Shelley E. Eating, Ulrich Wisloff, and Jeff S. Coombes. "Accuracy of Heart Rate Watches: Implications for Weight Management." *PLoS One* 11, no. 5 (2016): e0154420.
Wallhagen, Margaret. "The Stigma of Hearing Loss." *Gerontologist* 50, no. 1 (2010): 66–75.
Wamsley, Dillon, and Benjamin Chin-Yee. "COVID-19, Digital Health Technology and the Politics of the Unprecedented." *Big Data & Society* 8, no. 1 (2021): 1–6.
Waniata, Ryan. "Here One Review." *Digital Trends*, January 1, 2017. www.digitaltrends.com/headphone-reviews/Doppler-labs-here-one-review.
Weinstein, Adam. "Disney World Creepily Tracks Visitors NSA-Style with Magic Wristbands." *Gawker* (blog). January 2, 2014. http://gawker.com/disney-world-can-creepily-track-visitors-nsa-style-with-1493082046.
Welch, Chris. "Apple and Stanford's Apple Watch Study Identified Irregular Heartbeats in over 2,000 Patients." *Verge*, March 16, 2019. www.theverge.com/2019/3/16/18268559/stanford-apple-heart-study-results-apple-watch.

Weller, Chris. "Employees at a Dozen Fortune 500 Companies Wear Digital Badges That Watch and Listen to Their Every Move." *Business Insider*, October 21, 2016. www.businessinsider.com/humanyze-badges-watch-and-listen-employees-2016-10.

Wetsman, Nicole. "Light Sensors on Wearables Struggle with Dark Skin and Obesity." *Verge*, January 21, 2022. www.theverge.com/2022/1/21/22893133/apple-fitbit-heart-rate-sensor-skin-tone-obesity.

"Wheelchair User: Push/Step Counting." Fitbit Community. Accessed December 1, 2019. https://community.fitbit.com/t5/Other-Inspire-Trackers/Wheelchair-User-push-step-counting/td-p/3337961.

Whim TechNews. "Apple Pay—Official Announcement." Posted September 16, 2014. YouTube video. https://youtu.be/jiqSZRKskmk.

Whitson, Jennifer R. "Gaming the Quantified Self." *Surveillance & Society* 11, nos. 1–2 (2013): 163–176.

Williams, Raymond. *Culture and Society: 1780–1950*. New York: Anchor Books, 1960.

———. "Culture Is Ordinary." In *The Everyday Life Reader*, edited by Ben Highmore, 91–100. New York: Routledge, 2002.

———. *Keywords: A Vocabulary of Culture and Society*. Rev. ed.. New York: Oxford University Press, 1983.

———. *Marxism and Literature*. New York: Oxford University Press, 1977.

———. *The Politics of Modernism: Against the New Conformists*. New York: Verso, 1991.

Williamson, Ben. *Big Data in Education: The Digital Future of Learning, Policy, and Practice*. Thousand Oaks, CA: Sage Publications, 2017.

Winner, Langdon. *Autonomous Technology: Technics-Out-of-Control as a Theme in Political Thought*. Cambridge, MA: The MIT Press, 1977.

———. *The Whale and the Reactor: A Search for Limits in an Age of High Technology*. Chicago: University of Chicago Press, 1986.

Wolf, Gary. "The Data-Driven Life." *New York Times*, April 28, 2010. www.nytimes.com/2010/05/02/magazine/02self-measurement-t.html.

———. "Know Thyself: Tracking Every Facet of Life, from Sleep to Mood to Pain, 24/7/375." *Wired*, June 22, 2009. www.wired.com/2009/06/lbnp-knowthyself.

Wood, Stacy E. "Police Body Cameras and Professional Responsibility: Public Records and Private Records." *Preservation, Digital Technology, and Culture* 46, no. 1 (2017): 41–51.

Woods, Heather Suzanne. "Asking More of Siri and Alexa: Feminine Persona in Service of Surveillance Capitalism." *Critical Studies in Media Communication* 35, no. 4 (2018): 334–349.

Wright, James E., II, and Andrea M. Headley. "Can Technology Work for Policing? Citizen Perceptions of Police-Body Worn Cameras." *American Review of Public Administration* 51, no. 1 (2021): 17–27.

"WVU Rockefeller Neuroscience Institute Announces Capability to Predict COVID-19 Related Symptoms up to Three Days in Advance." West Virginia

University School of Medicine. May 28, 2020. https://medicine.hsc.wvu.edu/news/story?headline=wvu-rockefeller-neuroscience-institute-announces-capability-to-predict-covid-19-related-symptoms-up.

Yeginsu, Ceylan. "If Workers Slack Off, the Wristband Will Know (and Amazon Has a Patent for It)." *New York Times*, February 1, 2018. www.nytimes.com/2018/02/01/technology/amazon-wristband-tracking-privacy.html.

Young, Robin. "Beyond Counting Footsteps: Wearable Tech That Measures How You Work." WBUR. November 9, 2015. www.wbur.org/hereandnow/2015/11/09/humanyze-measuring-work-habits.

Zuboff, Shoshana. *The Age of Surveillance Capitalism: The Fight for a Human Future at the New Frontier of Power*. New York: Public Affairs, 2019.

Zulli, Diana. "Capitalizing on the Look: Insights into the Glance, Attention Economy, and Instagram." *Critical Studies in Media Communication* 35, no. 2 (2018): 137–150.

INDEX

Founded in 1893,
UNIVERSITY OF CALIFORNIA PRESS
publishes bold, progressive books and journals
on topics in the arts, humanities, social sciences,
and natural sciences—with a focus on social
justice issues—that inspire thought and action
among readers worldwide.

The UC PRESS FOUNDATION
raises funds to uphold the press's vital role
as an independent, nonprofit publisher, and
receives philanthropic support from a wide
range of individuals and institutions—and from
committed readers like you. To learn more, visit
ucpress.edu/supportus.